Metal-Oxide Gas Sensors

Edited by Soumen Dhara
and Gorachand Dutta

Published in London, United Kingdom

Metal-Oxide Gas Sensors
http://dx.doi.org/10.5772/intechopen.100736
Edited by Soumen Dhara and Gorachand Dutta

Contributors
Manar Lo Dayekh, Saleem Azara Hussain, Gopalakrishnan Chandrasekaran, Subha Krishna Rao, Rajesh Kumar Rajagopal, José Trinidad Guillen-Bonilla, Héctor Guillen Bonilla, Maricela Jiménez Rodríguez, Alex Guillen Bonilla, Verónica María Rodríguez Betancourtt, Víctor Manuel Rangel Cobian, María Eugenia Sánchez Morales, Antonio Casillas Zamora, Tomasz Hejczyk, Wiesław Jakubik, Jarosław Tomas Wrotniak, Vivekanand Mishra, Rashmi, Sukriti

First published in London, United Kingdom, 2023 by IntechOpen
IntechOpen is the global imprint of INTECHOPEN LIMITED, registered in England and Wales, registration number: 11086078, 5 Princes Gate Court, London, SW7 2QJ, United Kingdom

British Library Cataloguing-in-Publication Data
A catalogue record for this book is available from the British Library

Additional hard and PDF copies can be obtained from orders@intechopen.com

Metal-Oxide Gas Sensors
Edited by Soumen Dhara and Gorachand Dutta
p. cm.
Print ISBN 978-1-80356-962-8
Online ISBN 978-1-80356-963-5
eBook (PDF) ISBN 978-1-80356-964-2

Meet the editors

Dr. Soumen Dhara is a senior-grade assistant professor at Kalinga Institute of Industrial Technology, India. He received his Ph.D. in physics from the Indian Institute of Technology, Guwahati, India. He then obtained a prestigious JSPS postdoctoral fellowship at Kobe University from the government of Japan. He continued his research as a senior postdoctoral fellow at a number of institutions, including S. N. Bose National Centre for Basic Sciences, India, the University of Cambridge and Cardiff University, UK, and the Indian Institute of Science, Bangalore. He has more than 15 years of teaching and research experience and has published over 36 research articles and a review article in high-profile international journals (SCI indexed) as well as three book chapters. He has edited two international peer-reviewed book projects. His research interests are semiconductor nanostructures (nanowires, thin films, heterostructures), optoelectronics, solid-state lighting and light sensors, and spectroscopy of nanomaterials including ultrafast laser spectroscopy and thin-film transistors.

Dr. Gorachand Dutta is an assistant professor at the School of Medical Science and Technology, Indian Institute of Technology Kharagpur. He obtained his Ph.D. in biosensors and electrochemistry from Pusan National University, South Korea, where he developed different classes of electrochemical sensors and studied the electrochemical properties of gold, platinum, and palladium-based metal electrodes. He completed postdoctoral fellowships in the Department of Mechanical Engineering, at Michigan State University, USA and the Department of Electronic and Electrical Engineering, at the University of Bath, UK. His research interests include the design and characterization of portable biosensors, biodevices and sensor interfaces for miniaturized systems, and biomedical applications for point-of-care testing. He has expertise in label-free multichannel electrochemical biosensors, electronically addressable biosensor arrays, aptamer- and DNA-based sensors and surface bio-functionalization.

Contents

Preface

We are living in a society where automation with minimal or no human intervention is in great demand for all electrical devices and appliances. One of the major requirements for these smart devices is highly sensitive, long-lasting, low-cost smart sensors. A sensor is an analytical device that detects changes in the environment and converts them into a measurable analog resistance/voltage/current that can be read on a display or transmitted for further processing.

The healthcare industry demands low-cost, lab-on-chip type biosensors for simple and rapid detection of biomolecules or biogases. In the last two decades, a significant amount of research has been devoted to the development of various types of gas sensors using different nanomaterials. This book, written by scientists and researchers in the relevant professions, provides an overview of recent advances in the development of various gas sensors for the electronics and healthcare industries.

There are five chapters covering sensor basics, applications and numerical analysis. Dayekh et al. present an overview of gas-sensing technology. Mishra et al. discuss highly efficient optical-based gas sensors. Rao et al. review fiber-optic sensors and describe the use of spin-frustrated multiferroics on the surface of the fibers for the detection of volatile organic compounds. The detection of toxic gases using a metal oxide-based sensor and a commercial design is explained by Guillen Bonilla et al. In the final chapter, Hejczyk et al. describe a numerical analysis optimizing the steady-state condition of surface acoustic wave-based gas sensors.

The editors hope that this book will contribute to the development of new efficient and cost-effective gas sensors.

Soumen Dhara
School of Applied Sciences,
Kalinga Institute of Industrial Technology,
Bhubaneswar, India

Gorachand Dutta
School of Medical Science and Technology,
Indian Institute of Technology Kharagpur,
Kharagpur, India

Chapter 1

Gas Sensor and Sensitivity

Manar Lo Dayekh and Saleem Azara Hussain

Abstract

Gas sensors help to detect toxic and flammable gases in the atmosphere, and the use of these devices can reduce or prevent severe consequences for people and the environment. Metal oxides are one of the best materials used in the preparation of gas sensors, and they have proven in general that they have resistance to high temperatures Also, they are characterized by optical transparency at visible wavelengths, and they have a wide band gap. Whereas the interactive properties of metal oxides are the applications key chemical sensor. One of the characteristics of sensitivity is sensitivity, which is defined as the rate of change in the resistance of the thin film in the presence and absence of gas. Sensitivity is affected by several factors, including the relative humidity, the temperature of the sensors, the response time of the sensitivity, the time of exposure of the membranes to the gas, the background of the gas composition, and the thickness of the film.The chapter includes an explanation of the sensor parts and measurement sensitivity.

Keywords: gas sensing, sensitivity, gas sensitive materials, factors on gas, (Tio_2/rGo) nanocomposite, NH_3 gas sensor

1. Introduction

According to the scientific and technological developments, researchers has witnessed big dealing in the study of the applications of sensor including gas sensors and for its fundamental applications in the various fields of life like industrial industries, power generation, food, and beverages, medical, and therapeutic as well as agricultural industries. As a result of current industries possesses that increasingly involve the use and manufacture of highly hazardous materials especially toxic gases and flammable create a potential hazard to industrial facilities and their employees even the people who live near them. Moreover, all events that take place in the world represent as reminder for this problem [1].

2. Classification of gas-sensitive materials

According to the base of gas sensing, gas-sensitive materials are divided into two kinds based on electrochemical components and other principles, as shown in **Figure 1** [2].

There are three materials utilized as sensing components: metal oxide semiconductor, conductive polymer composites, and carbon nanomaterial [2].

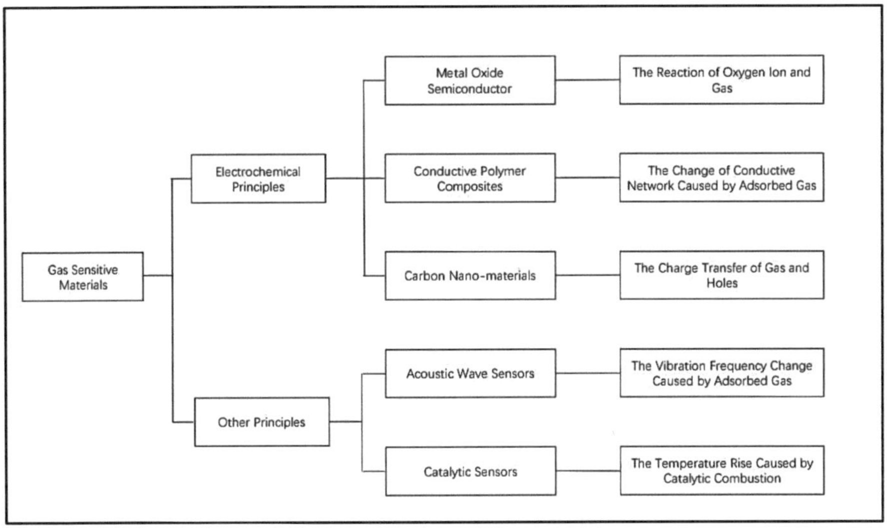

Figure 1.
Classification of gas-sensitive materials [2].

Metal oxides are among the best materials used in the preparation of gas sensors as shown in **Figure 2** and it proves generally that it resists high temperature. It is optically transparent at visible wavelengths and has a wide band gap [1, 3]. Both thin film and bulk semiconducting metal oxide materials have been widely used for the detection of a wide range of chemicals such as H_2, CO, NO_2, H_2S, ethanol, acetone, and human breath [4].

The reactive properties of metal oxide are the key to chemical sensing applications and when exposed to oxidizing gases including (O2) and gas (NO2) where the common denominator in the reactions of these gases is that they tend to form oxygen ions (O^-) or (O_2^-) that electrically active in order for the oxygen ion to be stabilized, it needs to diffuse into the vacant levels formed as a result of crystal defects within the composition of the substance in the following equations [3].

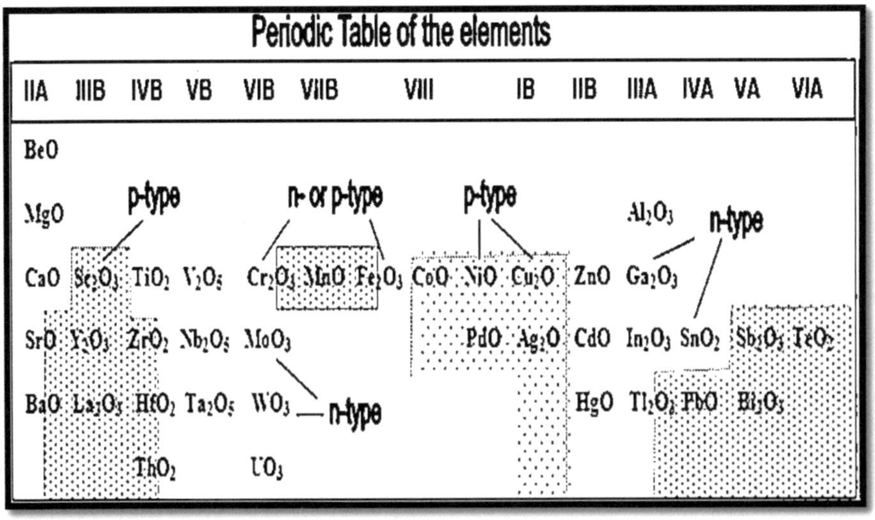

Figure 2.
Metal oxides are used in sensors [1].

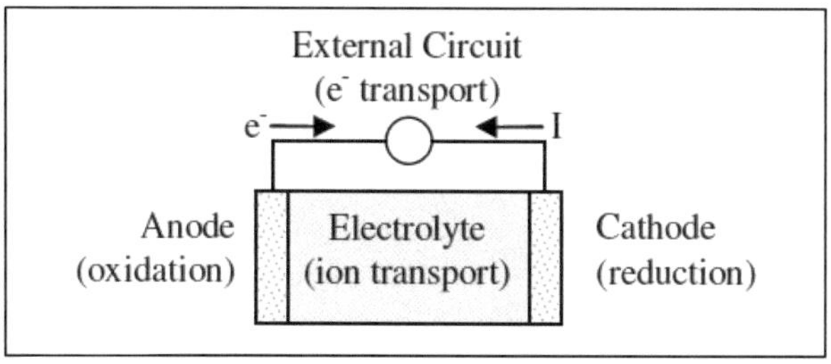

Figure 3.
Components of an electrochemical cell [4].

$$O_2 + e^- \rightarrow O^-_2 \quad (1)$$

$$O^-_2 + e^- \rightarrow 2O^- \quad (2)$$

Electrochemical cells contain individual electrodes. They are sites of chemical reactions that involve electron transfer with charged species (ions) that are transferred by electric current as it is showing in **Figure 3** where the anode is the electrode where the oxidation reaction takes place (the loss of electrons to the electrical circuit) while the cathode takes the results of the reduction reaction (gaining electrons from electric circuit) where electrical circuits permit the balance of the charge by electron transfer. It also provides the necessary electrical voltage and where the increase and decrease in resistance begin when exposed to one of the gases; therefore, the chemical sensors depend on the change in the resistance. The semiconductors are from the type (n-type) give a change in the resistance from the highest value to the lowest value if there is gas while (p-type) the change in resistance is in opposite way (from the lowest value to the highest value) in addition to the impurities here play a vital role in improving the interactive properties of the sensor modifying the reaction pathway sensitization and selectivity and increasing the detection limits of gas. Adding some kind of impurities is a catalyst and activator for these properties in addition to particle size and surface porosity are all activating factors to improve the sensitivity of gas and it is one of the necessary principles that must be noticed at the beginning of choosing materials for the chemical sensors [4].

Recent studies showed that it could improve the properties of sensors including selectivity and others by using a mixture of oxides of nanomaterials. The properties of the sensor are influenced greatly by granular size and structure porosity of materials. The particle boundary can be increased by decreasing the particle size and that lead to increase in the sensitivity and reduce the operating temperature which saves energy as well as recent researchers have turned to materials with nano-crystalline structure since it increases the improvement of sensors properties like eclectic and response time and this is achieved by providing a massive increase in the surface area [5]. **Figure 4** shows the gas component system.

3. Sensing properties

1. Sensitivity: Sensitivity can be defined as the percentage change in the resistance of the thin film in the existence of gas and lack thereof. Sensitivity is affected by factors including relative humidity, temperature of the sensors, the response time of the sensitivity, the time of exposure of the films to the gas composition,

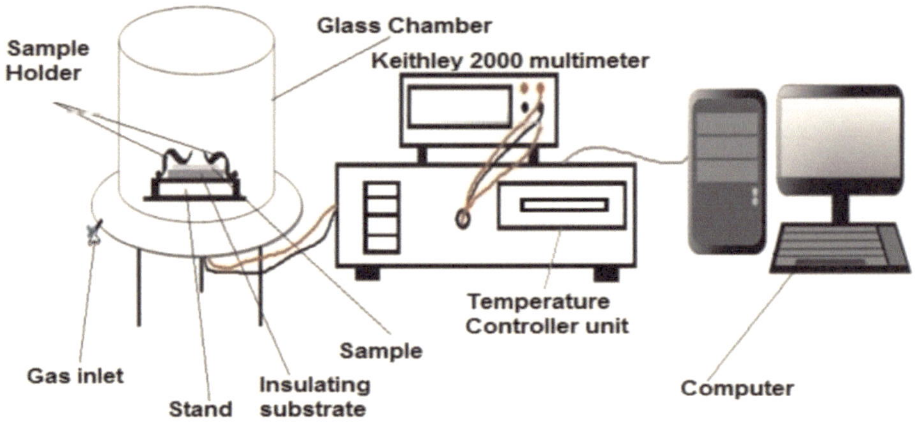

Figure 4.
Gas sensing system [6].

and the thickness of the film are symbolized by the symbol for (S) and can be shown in the following relation [1, 7, 8].

$$S = \frac{|\Delta R|}{R_o} \times 100\% = \frac{R_{gas} - R_{air}}{R_{air}} \times 100\% \tag{3}$$

Where:

S: Sensitivity.

(R_{gas}): Electrical resistance.

(R_{air}): Resistance in dry air.

(R_o): Resistance when entering the analytical gas

2. **Selectivity:** Selectivity can be defined as the ability of the sensor on identifying a specific gas inside a mixture of gases where the metal oxides have more sensitivity to the types of chemical gases in the air; therefore, studies underway to increase the selectivity of oxide semiconductor sensors.

3. **Response time and recovery time**: It is intended as the time the sensor takes for response in the case of changing the gas from zero to a specific value and reaches a stable value of about 90% from the final value. It is also defined as the time of recovery. It is time for the sensor signal to return to its initial value after removing the gas flow and reaching 10% [9].

4. **Stability**: Is the ability of the sensor to stabilize for a specific time while maintaining the selectivity, sensitivity, response time, and recovery time [10].

5. **Detection limit**:It is the minimum concentration of analysis that can be detected by the sensor.

6. **Dynamic range**:It is the range of concentration of examines between the maximum limit and the detection limit.

7. **Resolution**:It is the lowest focus that the sensor can detect.

8. **Linearity**:Is the relative deviation of the experimental graph on the ideal straight line in the ideal parameters.

9. **Hysteresis**:It is the large difference in the value of the output when the value converges with the decrease and increase in the degree of focus [11].

4. The effective factors on gas sensor properties

The following are the effective factors on gas sensor properties.

1. **Porosity**: The porous structure increases the reaction area, which increases the speed of response and improves the sensitivity and it helps to activate the molecules of gas on the surface of the semiconductor as the pores in the membranes act as channels to transport gas molecules which causes an increase in the response and retrieval speed [12]. The researcher Nayel et al. studied manufactured from a mixture of silver oxide and indium of the porous silicon slides, the results of the sensitivity tests showed that the sensor gave a high sensitivity at all temperatures to the base, where the sensitivity values ranged from single indium oxide, due to the effect of porous silicon [13].

2. **Grain Size**: The basic process that is related to the surface is adsorption and one of the effective factors in the adsorption on the sensor's surface is the grain size for the material and surface area. When grain size decreases, that will increase from the surface area leading to improve gas sensor properties. This is where sensors show high sensitivity towards gases. So if the grain size for the sensor membrane has a nano scale size, as the ratio of the surface to volume increases. Thus most of the studies tend to manufacture sensors of semiconductors membrane (nano). The decrease in grain size leads to increase in grain boundaries to increase the adsorption surface area to increase interaction resulting in increasing response speed and sensitivity value [14]. The effect of particle size on the sensitivity of chemically resistant nano-gas sensors was studied by the researcher Rothschild et al. by means of the effective carrier concentration as a function of the surface-state density of the SnO_2 sensor material, with different particle sizes ranging from 5, 80 nm [15].

3. **Dopant and Mixture**: It can be defined as adding impurities from a certain material to the pure semiconductors. These impurities act on controlling of optical, electrical, and constructive properties of the semiconductors. When adding impurities from certain materials to semiconductors, it works on the appearance of one type of charge carrier, which is called the majority charge carrier, and the diminishing of the other type, which is called the minority charge carrier. This led to the increase in electricity supply. The increase of impurities ratio works on decrease in grain size and this leads to the increase of surface area and grain boundaries and this helps to increase the interaction whereas gas

displaces the oxygen atoms which is related to the grain. This lowers the voltage barrier and thus lowers the resistance and conductivity increases. Since dopant increases the number of one type of carrier so the electrical conductivity depends on the number of atoms of the added material that is, focuses on free electrons generated from it [16]. The researcher Navaneethan M, et al., was able to study the improvement of sensitivity to ammonia gas sensor, as the sensitivity measurements for ammonia gas showed the highest response rate when Ag impregnation of ammonia gas [17].

4. **Junction Effect (P-N)**: When mixing two or more oxides to make a gas sensor, differential-difference will be created between them, due to the different affinity and work function p-n or p–p, or n-n was the type of electronic conductivity for those oxides. Thus, the number of charge carriers increases across the junction which leads to increase or decrease the electrical resistance for the sensor which is due to the contact of oxide p-n [16].

5. **Effect of Substrate**: The base directly affects the sensitivity properties of the prepared films usually using different types of substrate on which films are deposited such as silicon, quartz, glass, and aluminum. The porous silicon to work on improving sensor properties. Thus, the use of silicon substrate to precipitate films on it gives special different properties to oxides of semi-conductivity materials due to its influences on surface electric charges. Then, its effects on gas interaction with the sensor makes it required in the industry of gas sensors. The silicon etching process is one of the used techniques in the improvement of sensor properties due to the increase of surface area where it determines the regulation of pore size. The glass substrate is known for its optical properties as it has an absorption approach to zero and high transmittance within the visible area so it is used as a substrate to study the optical properties, constructive, and the sensitivity to the prepared film. Also, the proposed silicon acts to improve the sensor [18]. The researcher Rahman et al. studied the sensitivity of the engrafted nano silver oxide prepared by the solution method. The substance was deposited on glass carbon electrodes to give sensitivity with a quick response to methanol in the liquid phase and it gave good and stable allergic results.

6. **Active surface area and types of nanostructures**: In many gas sensors, the conductivity response is decided by the efficiency of catalytic reactions with detected gas participation, taking place at the surface of the gas-sensing material. As a result, control of the catalytic activity of gas sensor material is one of the most generally used means to improve the performances of gas sensors. But, in practice, the widely used gas-sensing metal oxide materials such as TiO_2, ZnO, Ga_2O3, SnO_2, Cu_2O, and Fe_2O_3 are the least active with the catalytic point of view. The pure SnO_2 thin film without any catalyst shows a very poor sensitivity ($\sim$3) confirming this statement [12]. Tian et al. presented a study of the gas sensing properties of Tio2 nanostructured materials and the effect of different forms of titanium dioxide on their sensitivity to gas. It was obtained that TiO_2 nanostructured materials with different dimensions show unique sensing properties for the detection of gases, in addition to that the ultra-small nanostructured materials with large and specific surface areas achieve rapid response for the detection of various gases [19].

5. Experimental gas sensor

5.1 Masks preparation and electrodes deposition

Preparation of the masks used for depositing the electrodes in the gas sensor, where these masks are made of steel material in the form of a single clip. The space between every two fingers of the clamp is ($400\mu m$) and the width of the fingers of the clamp is ($400\mu m$). Each piece includes five fingers its length is ($3.3mm$) and there is an active distance after the preparation of the masks, it is installed on the front part of the floors on which the material films are deposited to deposit the electrodes on top of it by thermal evaporation under vacuum conditions using a purity aluminum wire (99.9%) where the sedimentation takes place under pressure is shown in **Figure 5.**

5.2 Gas sensor preparation electrochemistry

The sensor can be prepared through the following steps:

a. Prepare the suitable floors (glass, Silicon)

b. Material deposition substances.

c. Deposition of the appropriate electrodes to make the sensor by thermal evaporation method in a vacuum.

d. Take the necessary measurements to determine the sensitivity of the material by using gas (NO_2, NH_4, … …..etc.).

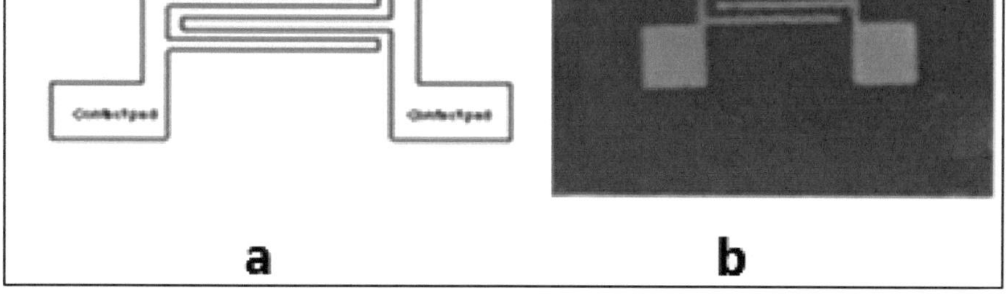

Figure 5.
(a) Aluminum mask planner for sensor gas. (b) the sample after the deposition of aluminum electrodes for gas sensor.

5.3 Description of the sensor system

The gas sensing chamber is a (cylinder or chamber) made of stainless steel, the cylinder is (30 cm) in diameter and (35 cm) high. The chamber contains several openings, including the gas pumping hole, the unloading hole, and the window opening to monitor what is going on inside the chamber, and a hole lead through is used for electrical connection between the parts inside the vacuum chamber and the measuring devices outside the vacuum chamber.

The sensing system contains a needle valve that controls the entry and exit of gas and is connected to a tube to a flask containing a source of gas, which supplies the gas to be tested in the chamber.

One of the important parts of the system is a laptop computer that is used to record the difference from sensors when exposed to a percentage of gas whose sensitivity is to be measured by the prepared membranes.

One of the parts of the sensing system is the hot plate, which is a base on which samples are placed inside the sensing chamber, the purpose of which is to raise the temperature of the membrane, as shown in **Figure 6**, and in order to control the operating temperature, the sensors are connected to a digital meter (thermometer).

The resistance of the membrane in the air is measured first, then the samples are entered into a vacuum chamber, and the resistance is measured as a function of time. The gas is pumped in and the change in the resistance of the membrane with time is read for every second with constant temperature, and readings are taken for both the pure and tainted cobalt films.

The gas sensing characteristics of (Tio_2/ rGo) nanocomposite to NH_3 sensor application has been studied by Ref. 17, and we have reached that Ammonia sensing acts for (Tio_2/ rGo) layered film tackled . Sensing behavior of the (Tio_2/ rGo) nanocomposites tested under the 100 ppm of environment. The (Tio2/ rGo) nanocomposite explained a high reaction to NH3 gas (48%) at room temperature.

It is clear in **Figures 7** and **8** that sensitivity (S) of NH3gas sensors based on (TiO_2/rGO) nanocomposite prepared by using pulsed laser ablation in Double distilled and deionized water (DDDW) at wavelength (532) nm to 100 ppm NH3gas in room-temperature. So that to obtain a measurement of sensitivity of the sample produced in this work, electrical resistance of nanocomposites was measured in the air and the presence of gas in room temperature. The resistive of gas sensors is called the relative change in resistance or conductivity for the nanocomposite. A known quantity of

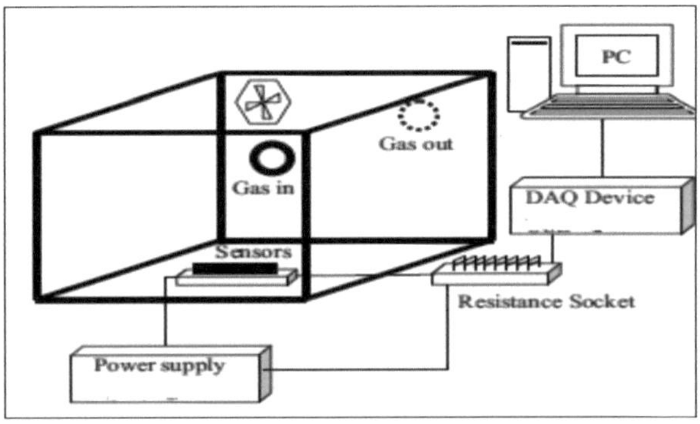

Figure 6.
The system used to measure sensitive membranes of gases [20].

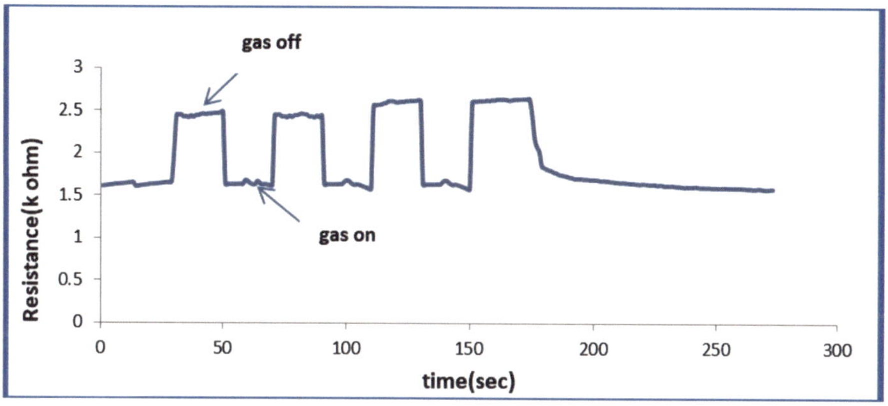

Figure 7.
Repetitive response curves of (TiO_2/rGO-1) nanocomposite exposed to 100 ppm NH3[21].

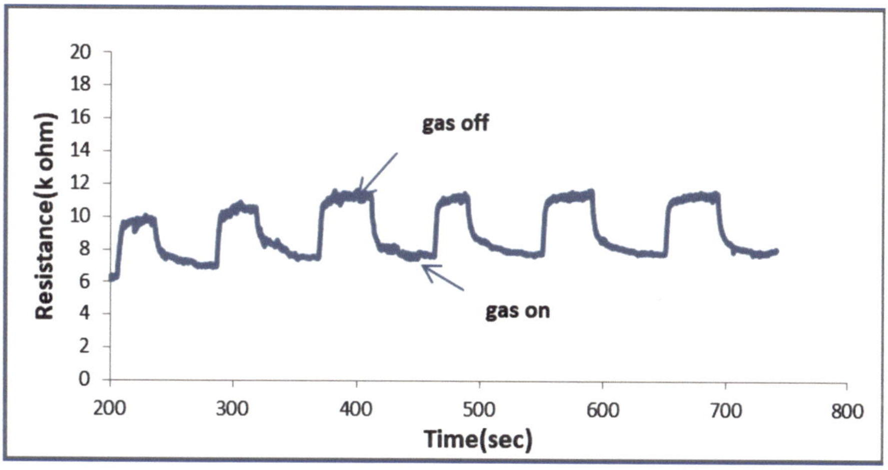

Figure 8.
Repetitive response curves of (TiO_2/rGO-2) nano composite exposed to 100 ppm NH3 [21].

intended gas is introduced after the ohmic strength of the sensor matter gets stability. The recovery features (as the target gas is withdrawn) are also controlled as a function of time. Sensitivity (S) can be calculated from Eq.(3). The sensitivity of TiO2/rGO-1 and TiO2/r GO-2 nanocomposites are (48) (25), respectively. The sensitivity of the (TiO_2/rGO-1) nanocomposite for ammonia gas is higher than that of (TiO2/rGO-2) nanocomposite because the sensitivity is based on grain size and grain boundary. If the distance between the grains is small, rising the interaction between oxygen absorbed and gases is rising, also the grain boundary will increase interaction and increase sensitivity, response, and recovery time is due to the first definition under exposure to NH3 at room temperature. It is necessary that both responses time and remedy time relied on gas focus and the temperature at which the sensor perform in this work. The operating temperature is constant (RT) and gas concentration is also constant and is estimated at 100 ppm, When exposed nanocomposites (TiO_2/rGO) to NH3 gas, the resistance of the nanocomposite decreases and thus make the sensor performance at room temperature, where Ammonia is an electron supplier and might assist electrons to the (TiO_2rGO) sensing matters at the sensation actions.

Semiconductor gas sensors are generally employed at the pressure in the atmosphere. Therefore, atmospheric oxygen on the surface adsorbs electrons from the conduction band of n-type (TiO_2rGO) nanocomposite film, forming O_2-and an electron-depleted layer at the surface of the film. NH3 gas was adsorbed, and electrons freed into the conduction band due to Equation below [4], decreasing resistance.

$$4NH_{3(g)} + 3O^{-}_{2(ads)} \rightarrow 2N_2 + 6H_2O + 6e^{-} \quad (4)$$

6. Conclusion

This chapter shows the importance of the study of gas sensors and their fundamental applications that are used in different aspects of life in addition to their importance in the dictation of toxic gases. The use of these devices can reduce the risks. To that, it is turn out the importance of classifying the materials that are used in gas sensitivity and components of gas sensors as it is explained the steps of sensitivity measurement with a description of the system used to sense the gases.

Author details

Manar Lo Dayekh* and Saleem Azara Hussain
University of Al -Qadisiyah, Diwaniyah, Iraq

*Address all correspondence to: manar_physics@yahoo.com

References

[1] Carpenter MA, Mathur S, Kolmakov A, editors. Metal Oxide Nanomaterials for Chemical Sensors. New York: Springer Science & Business Media; 2012

[2] Feng S, Farha F, Li Q, Wan Y, Xu Y, Zhang T, et al. Review on smart gas sensing technology. Sensors. 2019; **19**(17):3760

[3] Zhuiykov S. Electrochemistry of Zirconia Gas Sensors. Boca Raton: CRC Press; 2007

[4] Rheaume JM. Solid State Electrochemical Sensors for Nitrogen Oxide (NOx) Detection in Lean Exhaust Gases. Berkeley: University of California; 2010

[5] Zakrzewska K. Mixed oxides as gas sensors. Thin Solid Films. 2001;**391**(2): 229-238

[6] Patil PT, Anwane RS, Kondawar SB. Development of electrospun polyaniline/ ZnO composite nanofibers for LPG sensing. Procedia Materials Science. 2015;**10**:195-204

[7] Deshmukh S, Bari R. Nanostructured ZrO2 thin films deposited by spray pyrolysis techniques for ammonia gas sensing application. International Letters of Chemistry, Physics and Astronomy. 2015;**56**:120-130

[8] Korotcenkov G. Handbook of Gas Sensor Materials. Vol. 1. New York: Conventional Approaches; 2013

[9] Huang J, Wan Q. Gas sensors based on semiconducting metal oxide one-dimensional nanostructures. Sensors. 2009;**9**(12):9903-9924

[10] Scott RW, Yang SM, Chabanis G, Coombs N, Williams DE, Ozin GA. Tin dioxide opals and inverted opals: near-ideal microstructures for gas sensors. Advanced Materials. 2001;**13**(19): 1468-1472

[11] Gorley PM, Khomyak VV, Bilichuk SV, Orletsky IG, Horley PP, Grechko VO. SnO_2 films: Formation, electrical and optical properties. Materials Science and Engineering: B. 2005;**118**(1-3):160-163

[12] Wang C, Yin L, Zhang L, Xiang D, Gao R. Metal oxide gas sensors: Sensitivity and influencing factors. Sensors. 2010;**10**(3):2088-2106

[13] Nayel HH, AL-Jumaili HS. Fabrication of highly sensitive NH3 sensor based on mixed In2O3–AgxO Nanostructural Thin Films deposited on porous silicon. Journal of University of Anbar for Pure Science. 2019;**13**(2): 40-47

[14] Korotcenkov G, Cho BK, Gulina LB, Tolstoy VP. Gas sensing properties of SnO_2 thin films modified by Ag nanoclusters synthesized by SILD method. International Journal of Materials and Metallurgical Engineering. 2011;**5**(9):776-779

[15] Rothschild A, Komem Y. The effect of grain size on the sensitivity of nanocrystalline metal-oxide gas sensors. Journal of Applied Physics. 2004;**95**(11): 6374-6380

[16] Anand K, Kaur J, Singh RC, Thangaraj R. Preparation and characterization of Ag-doped In_2O_3 nanoparticles gas sensor. Chemical Physics Letters. 2017;**682**:140-146

[17] Navaneethan M, Patil VL, Ponnusamy S, Muthamizhchelvan C,

Kawasaki S, Patil PS, et al. Sensitivity enhancement of ammonia gas sensor based on Ag/ZnO flower and nanoellipsoids at low temperature. Sensors and Actuators B: Chemical. 2018;**255**:672-683

[18] Gründler P. Chemical Sensors: An Introduction for Scientists and Engineers. New York: Springer Science & Business Media; 2007

[19] Tian X, Cui X, Lai T, Ren J, Yang Z, Xiao M, et al. Gas sensors based on TiO_2 nanostructured materials for the detection of hazardous gases: A review. Nano Materials Science. 2021;**7**:396

[20] Su PG, Yang LY. NH3 gas sensor based on Pd/SnO2/RGO ternary composite operated at room-temperature. Sensors and Actuators B: Chemical. 2016;**223**:202-208

[21] Alnayli RS, Khadayeir AA, Dayekh ML, Hakim Z. Study of electrical and gas sensing characteristics of (TiO_2/rGO) nanocomposite to NH_3 sensor application. Journal of Physics: Conference Series. 2020;**1660**(1):012051

Chapter 2

Optical Gas Sensors

Vivekanand Mishra, Rashmi and Sukriti

Abstract

Miniature and highly efficient optical-based gas sensors have gained enormous consideration over the last few years. Materials based on the group-IV elements, namely silicon, germanium and their compounds, are deemed to be the potential candidates for the optical gas sensors. Optical gas sensors based on these materials offer appreciable sensitivity and high-density integration. Basically, these sensors paved the path for the flexible applications areas, namely internet of things (IoT), point-of-care testing, information and communication technology, etc. because of their potential candidature for being integrated with the several other photonic or electronic devices for on-chip signal processing and communication. Herein, we review optical gas sensors and discuss their basic principles, applications, recent advancement in the devices, etc. Gas concentrations can be easily detected and measured utilizing the characteristic optical absorption of gas species. This detection is crucial both for interpretation and observing of a wider range of phenomena extended from industrial practices to overall environmental change. Based on the findings, this review extends over a comprehensive overview of plethora of individual gas detection techniques, namely non-dispersive infrared, spectro-photometry, tunable diode laser spectroscopy and photo acoustic spectroscopy. This article focalizes over the discussion of the basic principle of the techniques introduced, their latest advancements and performance constraints, etc.

Keywords: sensors, optical gas sensors, spectrometers, infrared spectrometers, ultra-violet/visible spectrometers, fiber optic instruments, laser spectroscopy, optical and dielectric properties of gases

1. Introduction

The detection of gases is webbed across a plethora of application fields. Primary application extended over sector of petrochemical, where sensors were incorporated to safeguard (namely by detecting the leakage of highly toxic or flammable gases). The monitoring of feed stocks, as well as the key species parameters utilized in products as well as processes, keeps on revising depending upon the requirements with passage of time and day-to-day advancement in the field of science and technology [1, 2]. Several fields of science, namely atmospheric science, encourage the application of extremely sensitive gas detectors in order to detect, measure and develop the understanding of properties and trajectories of numerous gases, namely the greenhouse gases [3]. In the field of medical, investigation is made over several gases (namely nitric oxide (NO), ethane, ammonia (NH3), etc.) having potential candidature for

Materials	Advantage	Disadvantage	Target Gases and Application Fields
Metal Oxide Semiconductor	Lower Cost	Relatively lower sensitivity and selectivity	Industrial applications and civil use
	Shorter response time	Sensitive to environmental factors	
	Wider range of target gases	Higher energy consumption	
	Longer Lifetime		
Polymer	Highly sensitive	Long-time instability	Indoor air monitoring
	Shorter response time		Storage place of synthetic products as paints, wax or fuels
	Lower fabrication cost	Irreversibility	Industries namely, chemical industries
	Easy and portable structure	Poor selectivity	
	Lower consumption of energy		
Carbon Nanotubes	Ultra-sensitive	Complications in fabrication and repeatability	Partial discharge detection
	Great absorptive capacity	High cost	
	Large surface-area-to-volume ratio		
	Faster response time		
	Less weight		
Moisture Absorbing material	Lower cost	Vulnerable to friction	Humidity monitoring
	Less weight	Potential irreversibility in high humidity	
	Higher selectivity towards water vapor		
Optical Methods	High selectivity and stability, sensitivity	Difficulty in miniaturization	Remotely air quality monitoring
	Longer lifetime	Higher Cost	Gas leakage detection systems with excellent accuracy and safety
	Insensitive to environment change		High-end market applications
Calorimetric Methods	Stability at the ambient temperature	Danger of catalyst poisoning and explosion	Highly combustible gases under industrial environment
	Low cost	Intrinsic deficiencies in selectivity	Petrochemical plants
	Adequate sensitivity for industrial detection		Mine tunnels
			Kitchens
Gas Chromatograph	Outstanding separation performance	Higher cost	Laboratory analysis
	High sensitivity and selectivity	Difficulty in miniaturization for portable applications	

Materials	Advantage	Disadvantage	Target Gases and Application Fields
Acoustic Methods	Longer Life-time	Lower sensitivity	Components of the wireless sensor Networks
	Avoiding the secondary pollution	Sensitive to the environmental change	

Table 1.
Summary of the advantages, disadvantages and the myriad of applications for the listed methods of gas sensing.

biomarker gases application in fields of different for diagnostics, namely respiratory, etc. Chromatographs, analytical laboratory equipment has traditionally being utilized on larger scale for quantitative gas detection, but the equipment lacks the real-time data sampling [4]. Other than this sensor, semiconductor gas sensors or electrochemical devices were utilized. Other devices, namely Pellistorbeing, a potential sensor device (for the detection of the flammable gases closer to the lower range of the explosive limit but zero parts per million) for combustion on catalyst beads [5], semiconductor gas sensors can be very sensitive down to ppm [6], but they lack and suffer from the drift and erratic response to other gases as well as changing humidity levels. Other than this, electrochemical gas sensor being highly sensitive (ppm or ppb levels) and relatively specific to an individual gases suffers limited lifetimes and cross-response issues [7]. In comparative to this (**Table 1**), carefully designed optical absorption-based gas sensors are enriched with highly sensitivity, responsivity (approximately, 1 s), negligible drift, high-level gas specificity and seize to any cross response to other gases . They are blessed with highly controlled processes since the measurement corresponding to the detection can be made in real time and in situ devoid of any disturbance to sample under investigation [8].

Optical gas sensors act as a bridge among the sensors available at cheaper rates (with substandard performance) and costlier laboratory equipments. Optical gas sensors involve the transduction method which through direct measurement of molecules physical properties (e.g. the absorption of molecule at a particular wavelength) causes reduction in the drift. Since the intensity of the incident beam can be determined, the measurements are self-referenced and are compelled to be inherently reliable. Gas detection applications are webbed over the broader range of gas concentrations (proportion present in the air (or the other matrix) by the total volume). At standard temperature and pressure, gases mostly behave as an ideal gas and are approximately equal to molar concentration present in the given matrix. The study of widely used gas sensing methods based on the optical absorption measurements at particular wavelengths is the main focus of this article. Being non-dispersive gas sensing methods utilizing spectro-photometry, non-dispersive infrared, tunable diode laser spectroscopy and photo-acoustic spectroscopy [9–11].

Optical gas sensing proved to be the uncomplicated method and accomplish superior sensitivity, stability and selectivity over all other non-optical methods with much longer lifetime. Optical-based gas sensing has relatively shorter response time, enabling the detection on-line in real time. The performance of these sensors is immune to the any environment changes or catalyst poisoning triggered by the presence of specific gasses, etc. Optical gas sensing is spectroscopic investigation, but their applications on gas sensors are hampered because of miniaturization as well as comparatively high cost. That is the reason why only limited commercially available gas sensors based on the optical principle are available. Present review focusses over the general overview of optical-based gas sensors.

2. Classification of gas sensors

Factors affecting gas sensing mechanism:

Several factors should be taken into account when assessing the effectiveness of gas sensing techniques (**Figure 1** depicts the classification of gas sensor):

1. Sensitivity is a measure of the lowest volume concentration of a specific gas at which it can be detected.

2. Selectivity refers to the characteristic that analyses the capacity of a gas sensor to isolate a particular gas from a mixture of gases.

3. Response time is the amount of time taken by a gas sensor when the concentration of a gas reached to a certain value with respect to the time that it compels the sensor to generates an alarming signal.

4. Stability is the characteristic that tells whether the given sensing material can reconcile to its initial state once the detection is over. This includes retaining the response time, recovery time, sensitivity and selectivity.

5. Repeatability is the property of the sensor to exhibit certain response value when exposed to a certain value of gas concentration.

6. Fabrication cost, adsorptive capacity and energy consumption.

The operating stability of gas sensors that are intended for the market must also be guaranteed; or they must display a highly stable as well as repeatable signal over time.

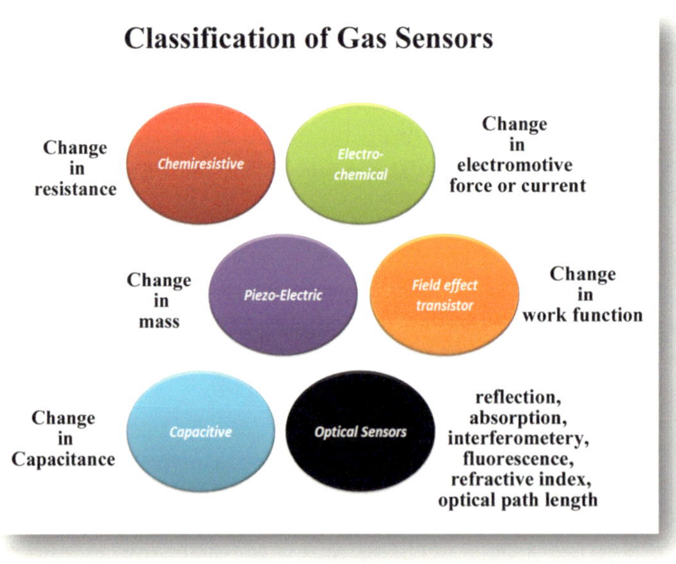

Figure 1.
Chart representing the classification of gas sensors.

But, the instability of the gas sensor is caused by a number of reasons such as design mistakes, structural modifications (such as changes in grain size or grain network), phase transitions (which typically refers to separation of the additives doped with sensing materials), poisoning brought on by the chemical processes as well as change in the environment.

The use of optical techniques for gas sensing is typically simple and can produce results with more sensitivity, stability as well as selectivity in respect to other non-optical techniques while having a significantly longer lifetime. Their quick response time makes on-line real-time detection possible. The changes in the atmosphere or poisoning of the catalyst brought on by particular gases, etc. will not affect performance. The majority of optical techniques utilized for sensing gas are based on spectroscopy.

2.1 Basic principles

There has been an extensive discussion about the principles of molecular absorption spectroscopy [12, 13]. Optical gas sensing techniques have been commercialized and while the majority are absorption-based, several other methods also play a significant role. Myriad of chemical substances show high absorption mainly in the region dominated over ultraviolet–visible range, near-infrared (NIR) as well as mid-infrared range of electromagnetic spectrum. Each substance has its unique set of absorption lines or bands that serve as the foundation for its detection and quantification. The parameters of the absorption spectra show variation in the various spectral bands displayed in **Table 2**.

Optical gas sensors work on the principle of the Beer and Lambert law and relation for the same is [14].

$$I(\lambda) = I_o(\lambda)e^{-\alpha c l} \quad (1)$$

were, $I(\lambda)\, and\, I_o$ are the detected as well as emitted optical intensity at a particular wavelength (λ), (respectively)

$\alpha(\lambda)[L/g]$= Absorption coefficient of the gas.

$c[g/L]$= Concentration of gas.

$l[m]$= beam-gas interaction path length.

Figure 2 depicts a sensor comprised of different parts, namely

i. Emitter (in order to generate the beam $(I_o(\lambda))$

ii. Optical Path (path length of gas cell (l) for guiding the beam interaction with gas.)

iii. Optical filter (for the selection of the characteristic wavelength range of the gas being target)

iv. Detector (for detecting the absorbed beam $(I(\lambda))$

The spectroscopic investigation relied mainly on the spectrometry absorption as well as emission. Absorption analysis is completely based on the Beer and Lambert

Spectral Range	Reason for the absorption
Ultraviolet (200–400 nm)	Electronic transitions
Near Infrared (700 nm to 2.5 μm)	Molecular vibrations and rotation, first harmonic
Mid Infrared (2.5–14 μm)	Molecular vibrations and rotation, fundamental

Table 2.
Origin of absorption spectra in the various regions of the electromagnetic spectrum.

law and is basically concentration-dependent absorption of the photons at a particular wavelength of the gas. Other than this plethora of improvised techniques of absorption spectrometry, namely Differential Optical Absorption Spectroscopy, Raman Light Detection, Tunable Diode Laser Absorption Spectroscopy, Differential Absorption LIDAR and Intra-Cavity Absorption Spectrometry (ICAS), etc. [15–17]. Spectrometry based on emission is observed when the atoms sitting in the excited levels on their transition to the ground state emits photons. Emission-based techniques are namely Laser-Induced Breakdown Spectroscopy, Fourier Transform Infrared Analysis (FT-IR) can be utilized for both the spectroscopy, namely absorption and emission. Furthermore, the photo-acoustic and correlation spectroscopy techniques also come in the category of spectroscopic analysis. These spectroscopic techniques are generally employed to gas detectors. This can contribute to the designing of the more complicated system at a higher cost level and can incorporate exceptional sensitivity, selectivity and reliability as compared to the other available gas sensors. Extensively utilized optical-based sensors utilize infrared source. The infrared source-based gas sensors work on the principle of absorption spectrometry. That implies that each and every gas is having a specific absorption property to infrared radiation of varying wavelength and results in unique fingerprint of infrared absorption. This implies that each and every gas has some unique property of absorption to infrared radiation range with varying wavelengths and thus shows a rare fingerprint of infrared absorption. Gas sensors based on infrared source consist of mainly three major parts, namely (depicted in **Figure 2**),

i. Infrared Source

ii. Gas chamber

iii. Infrared Detector

At the primary state, the infrared source emits broad-band of radiation wavelength containing the wavelength range corresponding to the absorption of target gas. After that the gas sample present in the gas cell absorbs the incoming radiation beam of their specific wavelength. Later on the optical filters are utilized to screen out the entire wavelength of the radiation except the radiation wavelength absorbed by the target gas. Thus, the gas of interest can be easily detected as well as measured just by employing infrared detector. The entire setup is named as Non-Dispersive Infrared gas sensor.

Non-dispersive sensing technique is utilized, wherein the beam (unfiltered) is utilized for the interaction with the gas. This sensing technique encourages the selective detection of beam (λ), just by filtering the beam detected by characteristic absorption spectra of the molecular species. Non-dispersive infrared (NDIR) sensors conventionally configured with infrared emitters and infrared detectors [18]. Sensors

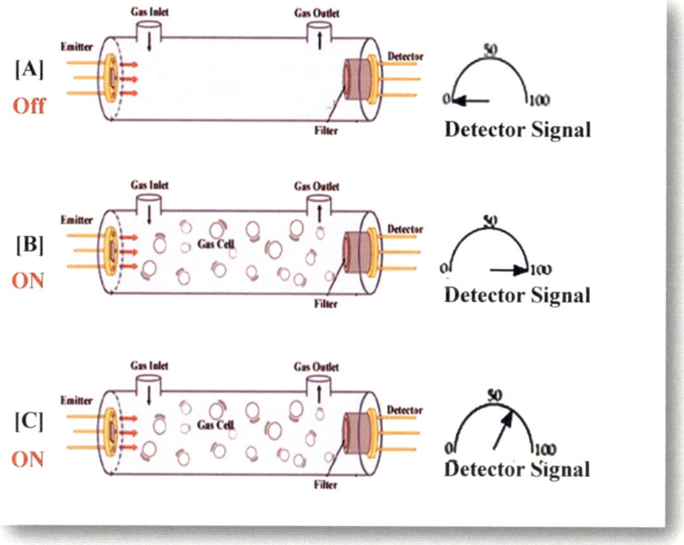

Figure 2.
Beer and Lambert law–based optical gas sensor; (A) off state (no any signal being detected, (B) on state maximum detected signal, (C) decreased detection with gas concentration.

work on the same principle of Beer–Lambert's Law, if utilized for different spectral regions or configured for acoustic detectors.

A different gas cell consisting of the reference gas other than the single detector mode can be incorporated to uplift the overall accuracy of the infrared sensing method just by truncating the ambient environmental factors without inducing much more complexity. **Figure 2** depicts the layout of two detectors. Just to keep the radiation parameters of infrared source identical, gas cell utilizes the reflected infrared beams from the single source just utilizing the mirror property. The assortment of the range of wavelength of the infrared beam utilized as source impacts a lot on the concluding detection result. Spectral region of the mid-infrared gains more attention as it encourages stronger molecular absorption as compared to spectral region of near-infrared. The conventional mid-infrared laser sources suffer major drawbacks, namely wavelength tunability, lower power of output, complexity, requirement of coolants. Plethora of lasers have been developed and investigated in order to overrule the above mentioned disadvantages, e.g. Quantum-Cascade Lasers blessed with tunability, highly narrow line-width as well as lower average power for the operation at the room temperature.

Myriad of topologies have been employed for the fabrication of optical gas sensors. The commonly utilized sensors are based on gas cells formed between face-to-face configured emitters and optical detectors. Several approaches followed for the miniaturization of the gas cell include utilization of enhancement layers namely, Utilization of photonic crystals [19], optical cavities [20], multi-pass cells [21], gas enrichment layers [22–24], in order to enhance beam-gas interaction, the planar configurations of the emitters as well as detectors, utilization of waveguides for the interaction of evanescent field. Optical detectors, namely photodiode, thermopile or pyroelectric, acoustic detector are mainly implemented in order to detect the beam being absorbed ($I(\lambda)$) [25–28]. The mechanism of lock-in detection is employed to extract the sensor response signal ($I(\lambda)$) and further known frequency is utilized to modulate the emitter. The reference detector is frequently utilized in order to compensate for the variation in emitted beam. Further, additional sensors can be utilized in order to compensate for the several parameters of environment, namely temperature, pressure, etc.

Types of Optical Gas Sensors

- Optical sensors' operation mainly includes controlling, then detection of the beam propagating through the target area where the photons which got detected generate the electrical signals. Frequently utilized optical sensors are, namely fiber-optic gas sensors and photonic-crystal gas sensors; these sensors are based on the principle of the detection of the beam propagation through utilization of devices. Fiber-optic-based sensor measures the modification in optical property (namely, wavelength) and detects the analytes introduced on the sensing polymeric layer by employing optical fiber. These sensors are enriched with excellent sensitivity, stability with respect to the environmental factor as well as longer lifetime. Implementation of optical fiber in the structure causes complications in the miniaturization. Structure of the photonic crystal sensor utilizes the periodic arrangements of dielectric materials with numerous refractive indexes.

- In order to detect the carbon monoxide, mid-infrared photonic crystals are commonly utilized. By the utilization of the advanced micro-machining technology, we can precisely control or tune the dimension as well as the shape of the pattern of the sensing device. But the implementation of this advance technique can lead to the enhancement in the overall fabrication price of the device. Optical gas sensors enriched with the detection level ranging over then ppm range have been successfully developed.

- Spectroscopy-based methods have traditionally included infrared absorption spectroscopy, gas chromatography (GC) and Raman scattering spectroscopy as the standard gas detection methods. Despite their sturdiness, the aforementioned spectroscopic sensing approaches typically need expensive and large apparatus, making them difficult to access and unsuitable for applications that call for in situ monitoring and on-site analysis. Therefore, alternate gas sensing methods are being developed concurrently for lower cost and greater portability.

- Electrical gas sensors such as chemiresistors, electrochemical sensors, Schottky diodes, field-effect transistors, chemiresistive sensors and impedance sensors are among these types of gas sensors.

- Optical gas sensors such as fiber-optic gas sensors and photonic crystal gas sensors are also available. Despite having been developed and commercialized gas sensors for a long time, traditional materials of gas sensing such as oxides of metals, conducting polymers and carbon nano tubes (CNTs) still face issues including low selectivity, high operating temperatures and poor repeatability. As a result, ongoing efforts are needed to explore and develop new gas sensing materials.

- Among the most used type of optical gas sensors are fiber gas sensors, as shown in the figure.

A waveguide with a fiber core and surrounding cladding makes up its primary part. The core allows a small amount of light to transmit in form of evanescent waves perpendicular to the axis of fiber with exponentially diminished intensity in the cladding. On the other hand, majority of the light travels along with the fiber core via total internal reflection because core has a greater refractive index than the cladding. The alteration in the refractive index of the cladding is caused by the adsorption of the molecules of gas on its surface, which in turn affects the wavelength and intensity of the output light signals. Due to the limited selectivity of the gas adsorption-induced variation in the overall refractive index of the material, the cladding is typically modified with molecules or nano-materials that have a particular affinity for the target gas.

3. Applications

A side-polished fiber (SPF) coated with cholesteric liquid crystal film (CLCF) has been employed for Volatile Organic compound (VOC) sensing, where a rise in the concentration of VOC on CLCF causes a rise in the pitch of the resulting light. As a result, the resonant dips exhibit a blue shift that can be linked to the exposure to VOCs. For acetone, tetrahydrofuran and methanol gas, the sensitivities of CLCF-SPF have been observed to be 3.46, 7.08 and 0.52 nm-L/mmol, respectively. It can be observed that the overall sensitivity of the CLCF-SPF rises with the VOCs' molar mass.

- ZnO nanoparticle-coated fiber-optic sensors have demonstrated the relation between concentration and selectivity for ammonia, acetone and ethanol. Due to the increment in the catalytic reactivity of the acetone at elevated concentrations, ZnO nano-particles exhibit exceptional sensitivity towards the gas ammonia and acetone at lower and higher concentrations of up to 150 ppm and above 150 ppm, respectively.

- A CO gas sensor that operates at room temperature and uses ZnO detecting film has been created. Sensor uses the mechanism of surface Plasmon resonance mechanism, which has been demonstrated as a possible contender for the commercial applications of the detection of CO. According to reports, this sensor has a wider range of CO concentration, webbed over the range of 0.5–100 ppm at the room temperature and a faster response time of roughly 1 s. It also has a higher sensitivity of about 0.091 ppm and has been demonstrated that these sensors are highly selective for CO, with barely any interference from any other gases, namely CO_2, NO_x, NH_3, LPG and H_2.

Figure 3 depicts photonic crystal gas sensor, as another type of optical gas sensor that is similar to one described above. The photonic crystals are mainly synthetic optical materials having periodic variations in value of the refractive index. According to photonic band gap theory, beams having the wavelength in range of photonic crystal band gap can be bound in and transmitted down the air channel having little energy loss in the case where the cladding of fiber has a higher refractive index compared to the photonic crystal core with air channels. Additionally, air passages act as the cells for the insertion of the molecules of the target gas, which alters core's refractive index of photonic crystal and hence, alters the output illumination.

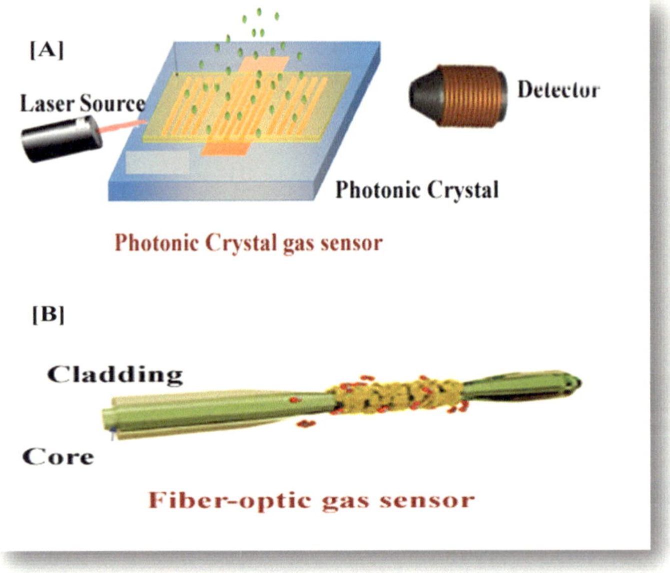

Figure 3.
Schematic illustration of (A) photonic crystal gas sensor and (B) fiber-optic gas sensor.

Three patterns of varying sizes on a single substrate have been used to identify different VOCs using Si-based photonic crystals. By using electrochemical anodization to create mesopores of various sizes, the designs are etched onto the Si wafer. On a Si substrate, 8 nm of diameter mesopores were created in multilayers with 178, 229 and 300 nm of the vertical spacing, allowing the refection of 430, 580 and 740 nm photons. Introduction of the analytes over this photonic crystal with several layer patterns raises the effective refractive index and changes the permitted wavelength of the reflected beam. When sensing capabilities of the devices were examined using methanol, ethanol, etc. in the nitrogen carrier gas presence, LOD in the ppm range was observed. Using this technique, analytes by calculating the shift in wavelength in time of each VOC.

Researchers have looked into self-assembling silica nanospheres to create photonic crystals for ethanol, water and carbon disulfide (CS2) detection. On silica colloid drying over substrate and then annealing it at the temperature of 600°C for the sintering, the silica photonic crystal can be created. To improve the interaction with analytes, HKUST-1 can be coated on the nano-structured silica. Each silica nano-sphere has a diameter of around 300 nm and is arranged in a face-centred cubic configuration. For a gas-sensing test, when analytes are present, NIR light can be shone in the direction of the silica nanostructure. The predicted detection limit for water, ethanol and CS2 is 2.6, 0.3 ppm and CS2 is 0.5 ppm, according to tests on the sensing ability of these substances.

4. Conclusion

To conclude, optical gas sensing remains an important field that complements other gas detection technologies. The opportunities afforded by new technology, together with the challenges that remain, will make this an exciting and rapidly developing field for many years to come. Commercially available gas sensors are rarely based on the optical principles, present article focuses over the broad overview of techniques involved in gas analysis and illustration of the detailed introduction of the optical-based gas sensors. Optical sensors including fiber-optic and photonic crystal gas sensors operate based on the detection of the light propagation through the device. Fiber-optic sensors detect analytes by measuring optical property changes of light such as wavelength, which is introduced on the polymeric sensing layer through optical fiber. These sensors have many advantages in high sensitivity, stability to environmental factors and long lifetime. However, needs of optical fiber in the structure make it difficult to be miniaturized. Photonic crystal sensors employ periodic arrangements of dielectric materials with various refractive indexes. Mid-infrared PhCs are a common example of these devices to detect common gases such as CO. These sensors can be developed using advanced micro-machining technology and, therefore, the dimension and shape of the device pattern can be precisely controlled, which on the other hand increases the fabrication cost.

Author details

Vivekanand Mishra[1]*, Rashmi[1] and Sukriti[2]

1 Division of Physics, Department of Science, Alliance University, Bengaluru, India

2 Department of Physics, National Institute of Technology, Kurukshetra, India

*Address all correspondence to: vive2009@gmail.com

References

[1] Hodgkinson J, Tatam RP. Optical gas sensing: A review. Measurement Science and Technology. 2013;**24**:012004

[2] Allen MG. Diode laser absorption sensors for gas-dynamic and combustion flows. Measurement Science and Technology. 1998;**9**:545-562

[3] Laj P et al. Measuring atmospheric composition change. Atmospheric Environment. 2009;**43**:5351-5414

[4] Smith D, Španěl P. The challenge of breath analysis for clinical diagnosis and therapeutic monitoring. Analyst. 2007;**132**:390-396

[5] Jones E. In: Moseley PT, Tofield BC, editors. The Pellistor Catalytic Gas Detector Solid State Gas Sensors. Bristol: Adam Hilger; 1987. chapter 2

[6] Williams DE. In: Moseley PT, Tofield BC, editors. Conduction and Gas Response of Semiconductor Gas Sensors Solid State Gas Sensors. Bristol: Adam Hilger; 1987. chapter 4

[7] Bakker E, Telting-Diaz M. Electrochemical sensors. Analytical Chemistry. 2002;**74**:2781-2800

[8] Lackner M. Tunable diode laser absorption spectroscopy (TDLAS) in the process industries—A review. Reviews in Chemical Engineer. 2007;**23**:65-147

[9] Crowder JG, Hardaway HR, Elliott CT. Mid-infrared gas detection using optically immersed, room-temperature, semiconductor devices. Measurement Science and Technology. 2002;**13**:882-884

[10] Rawlins WT, Hensley JM, Sonnenfroh DM, Oakes DB, Allen MG. Quantum cascade laser sensor for SO2 and SO3 for application to combustor exhaust streams. Applied Optics. 2005;**44**:6635-6643

[11] Varga A, Bozoki Z, Szakall M, Szabo G. Photoacoustic system for on-line process monitoring of hydrogen sulfide (H2S) concentration in natural gas streams. Applied Physics B. 2006;**85**:315-321

[12] Banwell CN, EM MC. Fundamentals of Molecular Spectroscopy. 4th ed. London: McGraw-Hill; 1994

[13] Ingle JD, Crouch SR. Spectrochemical Analysis. London: Prentice Hall; 1988

[14] Bernath PF. Spectra of Atoms and Molecules. 3rd ed. Oxford, UK; New York, NY, USA: Oxford University Press; 2016. p. 465

[15] Miao Y, Yao Q, Qiu N, Zhang J. Application research of laser gas detection technology in the analysis of Sulphur hexafluoride. In Proceedings of 2010 China International Conference on Electricity Distribution (CICED), Nanjing, China. 13-16 September 2010. pp. 1-3

[16] Shi K, Burris JF, Newchurch MJ, Johnson S, Long S. Differential absorption LIDAR to measure subhourly variation of tropospheric ozone profiles. IEEE Transactions on Geoscience and Remote Sensing. 2011;**49**:557-571

[17] Refaat TF, Ismail S, Koch GJ, Rubio M, Mack TL, Notari A, et al. Backscatter 2-μm lidar validation for atmospheric CO2 differential absorption lidar applications. IEEE Transactions on Geoscience and Remote Sensing. 2011;**49**:572-580

[18] Liu X, Cheng S, Liu H, Sha H, Zhang D, Ning H. A survey on gas sensing technology. Sensors. 2012;**12**:9635-9665. DOI: 10.3390/s120709635

[19] Kraeh C, Martinez-Hurtado JL, Popescu A, Hedler H, Finley JJ. Slow light enhanced gas sensing in photonic crystals. Optical Materials. 2018;**76**: 106-110

[20] Ayerden NP, de Graaf G, Enoksson P, Wolffenbuttel RF. A highly miniaturized NDIR methane sensor. In Proceedings of the SPIE, Micro-Optics, Brussels, Belgium; Vol. 9888. 3-7 April 2016. p. 98880D.

[21] Chen K, Zhang B, Liu S, Jin F, Guo M, Chen Y, et al. Highly sensitive photoacoustic gas sensor based on multiple reflections on the cell wall. Sensors and Actuators A Physical. 2019;**290**:119-124

[22] Moumen S, Raible I, Krauß A, Wöllenstein J. Infrared investigation of CO2 sorption by amine based materials for the development of a NDIR CO2 sensor. Sensors and Actuators B: Chemical. 2016;**236**:1083-1090

[23] Tan Q, Tang L, Yang M, Xue C, Zhang W, Liu J, et al. Three-gas detection system with IR optical sensor based on NDIR technology. Optics and Lasers in Engineering. 2015;**74**:103-108

[24] Tombez L, Zhang EJ, Orcutt JS, Kamlapurkar S, Green WMJ. Methane absorption spectroscopy on a silicon photonic chip. Optica. 2017;**4**:1322-1325

[25] Xing Y, Urasinska-Wojcik B, Gardner JW. Plasmonic enhanced CMOS non-dispersive infrared gas sensor for acetone and ammonia detection. In Proceedings of the 2018 IEEE International Instrumentation and Measurement Technology Conference (I2MTC), Houston, TX, USA. 14-17 May 2018. pp. 1-5

[26] Dong M, Zheng C, Miao S, Zhang Y, Du Q, Wang Y, et al. Development and measurements of a mid-infrared multi-gas sensor system for CO, CO2 and CH4 detection. Sensors. 2017;**17**:2221

[27] Sklorz A, Janßen S, Lang W. Detection limit improvement for NDIR ethylene gas detectors using passive approaches. Sensors and Actuators B: Chemical. 2012;**175**:246-254

[28] Dharmendra R, Mishra V. Implementation and analysis of 32 channel MDRZ DWDM system at narrow channel spacing. Journal of Physics Communications. 2022;**6**:095003, 1-7

Chapter 3

Fiber Optic Sensors for Gas Detection: An Overview on Spin Frustrated Multiferroics

Subha Krishna Rao, Rajesh Kumar Rajagopal and Gopalakrishnan Chandrasekaran

Abstract

Real-time gas sensors, which use chemiresistive metal oxide (MO) semiconductors, have become more important in both research and industry. Fiber optic metal oxide (MO) semiconductor sensors have so increased the utility and demand for optical sensors in a variety of military, industrial, and social applications. Fiber optic sensors' inherent benefits of lightweight, compact size, and low attenuation were actively leveraged to overcome their primary disadvantage of expensive cost. With the growing need for quicker, more precise, and simpler gas sensing, metal oxide semiconductor gas sensors are focusing on new and novel materials at room temperature. The realization that materials with coexisting magnetic and ferroelectric orders offer up effective ways to alter magnetism using electric fields has drawn scientists from diverse areas together to research multiferroics for gas sensing applications in recent years. The chapter shall encompass a brief summary of the underlying physics related to fiber optic gas sensors and parameters involved in gas sensing, the significance of the fascinating class of metal oxide materials, and an outline of spin frustrated multiferroics for possible applications and its potential possibilities for progress in the future.

Keywords: fiber optic sensors, Multiferroics, spin frustrated, $Bi_2Fe_4O_9$, $YMnO_3$, $Ni_3V_2O_8$

1. Introduction

The air we breathe contains a variety of chemical compounds, some of which are beneficial and others that are harmful. Toxic and hazardous gas emissions have become a source of worry in recent years [1]. As a result, there has been an increase in demand for gas detection and monitoring. Generally, a gas sensor should be designed to accomplish two critical functions: receptor and transducer. The capacity to detect certain gas species (via interactions such as adsorption, chemical, or electrochemical reaction) is the function of a receptor, while the ability to convert gas identification into a sensing signal is the function of a transducer.

Gas sensor research is designed to reinforce two most important functions: the receptor function, which is generally connected to selectivity, and the transducer function, which is directly related to the sensitivity of a gas sensor [2]. To fulfill the two purposes, different sensing techniques are applied, resulting in distinct sensor approaches such as catalytic gas sensors [3], electrochemical gas sensors [4], optical gas sensors [5], thermal conductivity gas sensors [6], and acoustic gas sensors [7, 8]. Over the last few decades, Optical gas sensors have been projected as a good choice considering their benefits compared to other available sensing technologies, as they are immune to electromagnetic interferences [9], do not require any electric power to work, and the possibility of multiplexation, ability to perform in remote areas and under harsh environmental conditions [10].

Commercial applications exist for gas detection using optical fibers. Owing to its advantages and good productivity over non-fiber sensors, such as metal oxide semiconductor (MOS) and spectroscopic approaches, optical fiber gas sensors are still being extensively examined. Between 2017 and 2023, the global optical sensing market is expected to grow at a CAGR of 15.47 percent, from USD 1.13 billion in 2016 to USD 3.47 billion by 2023 [11]. The physical qualities of optical sensors that

Types of Chemical sensors	Manufacturers commercially available	Range of detection / Concentration (ppm)	Response time (s)	Compounds
Electrochemical Sensor	Environmental sensors Co, Winsen, City tech	0–30 ppm	60 s	ethanol. formaldehyde
Metal Oxide Semiconductor Sensors	Aeroqual, Unietc SRL, Applied Sensor AMS,	10–5000 ppm	10–60 s	alcohols, aldehydes, amines, aromatic hydro-carbons (petrol vapors, LPG, etc.)
Non-dispersive Infrared Sensors (NDIR)	Wuhan Cubic, Alphasense	0–100%	< 45 s	IR absorbing VOC (methane)
Thermal Sensor (Pellistor)	Figaro, Microcel, Sixth sense	0–100%	10–40 s	Most combustile vapors (methane, iso-butane, propane)
Photoionization Sensor (PID)	Mocon Baseline series, Alphasense Gray wolf	10–20,000	< 3 s	VOC's with proper ionization potential (isobutylene, aromatic hydrocarbons)
Fiber Optic sensor	Oxsensis Ltd. (UK), RJC Enterprises LLC (US), Silixa Ltd. (UK), OPTEK Technology Inc. (US), Opsens Inc. (Canada), Intelligent Fiber Optic Systems Corporation (US) and Fotech Solutions Limited (UK).	0–500 ppm	< 30 s	Flammable VOCs such as ammonia, ethanol, acetone, methanol, benzene, IPA etc.

Table 1.
Varous types of chemical sensors for measuring volatile organic compounds that are commercially available.

make them suited for sensing in difficult settings, as well as continuous technological advancements in optical sensors, are major drivers of market growth. **Table 1** displays a comparison of the commercially available chemical sensors employed for the detection of Volatile organic compounds (VOCs). The manufacturer's primary goal is to develop a sensor with the descent range of detection and the lowest response time possible. The choice of the sensor basically depends upon the type of gas that has to be detected, concentration range, if the sensor is intended to be stationary or portable, detect the presence of other gases that could possibly damage the measuring device. A few commercially available sensors are Electrochemical (Amperometric) Sensors [12], Metal oxide semiconducting sensors [13–15], Nondispersive Infrared Sensors (NDIR) [16, 17], Photoionization Sensor (PID) [18–20], Thermal Sensor (Pellistor) [21, 22], Fiber optic sensors [23–28].

1.1 Fiber optics sensors (FOS)

After the advent of lasers in 1960, researchers became interested in studying the possibilities of optical fiber communication systems for sensing, data communications, and a variety of other applications [29]. As a result, fiber optic communication systems have become the preferred mode of data transfer for gigabits and beyond. This sort of fiber optic communication is used to send data, voice, telemetry, and video over long distances or within local area networks (LANs) [30, 31]. By converting electronic signals into light, this technique employs a light wave to convey data via a fiber. A few outstanding characteristics of this technology being light in weight, possess low attenuation, smaller diameter, long distance signal transmission, transmission security, etc. [32]. The recent improvements in fiber optic technology have had a significant impact on telecommunication technology. Designers combined the productive outcomes of optoelectronic devices with fiber-optic-telecommunication devices to produce fiber optic sensors in the last revolution. Many of the parts used in these devices were originally designed for fiber-optic sensor applications. Fiber optic sensors have surpassed traditional sensors in terms of capability [33]. The development of optical sensors has recently been described and discussed in a number of excellent review studies from various angles. Mahata et al. provided an overview of the use of rare-earth-based MOFs as luminous sensors to identify nitro explosives, cations, anions, small molecules, pH, and temperature [34]. On-chip biological sensors based on optofluidic photonic crystal cavities were reviewed in-depth by Zhao and his colleagues, and the sensing theories and uses for these sensors were covered in detail [35]. Yoon's group has made significant contributions to fluorescence-based optical sensors and offered a perceptive viewpoint on the advancement of fluorescent chemosensors [36–38].

1.1.1 Optical fibers—An overview

Optical fibers are light-transmitting waveguides with two primary components: a core made of glass and a cladding composed of a material with a lower refractive index than the core as shown in **Figure 1a**. The optical fiber is protected against physical damage and scattering losses produced by micro bending by an extra elastic layer as a buffer composed of plastic surrounding the cladding section. The jacket layer is the final layer, and it can be used to identify the fiber type. Because of its purity, quartz glass is used to make the majority of fibers [39]. Total internal reflection occurs at the interface between the core and the cladding in optical fibers as long as the angle

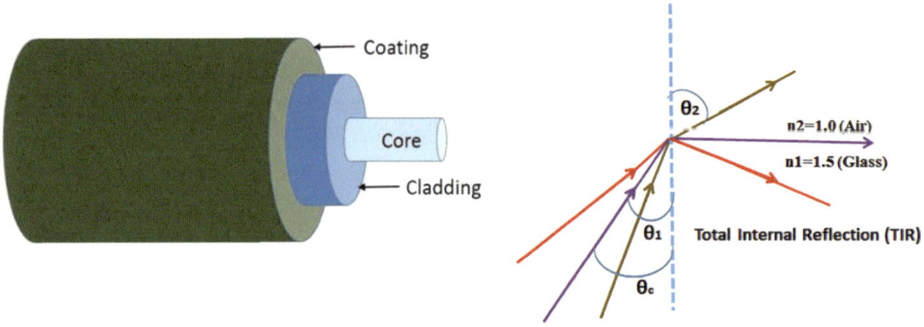

Figure 1.
(a) Basic components of optical fiber; and (b) principle of operation in fiber optics.

of incident light inside the core is greater than the critical angle. In this way, incident light is reflected back into the core and propagated through the fiber (**Figure 1b**). If the light strikes the interface at a greater angle than the critical angle, it will not pass through the opposite medium. The angle of incidence, as well as the core and cladding refractive indices, are all factors to consider. The amount of light reflected at the interface is determined by the angle of incidence as well as the refractive indices of the core and cladding [40].

Generally the number of modes and the refractive index are used to divide optical fiber into two categories.

1.1.1.1 Single mode fiber (based on number of modes)

Only one kind of light ray can pass through a single-mode fiber. This fiber has a tiny core diameter (5 μm) and a large cladding diameter (70um), with a minimal refractive index difference between the core and the cladding. There is no signal deterioration when travelling through the fiber because there is no dispersion. Using a laser diode, light is sent through it [41] (shown in **Figure 2a**).

1.1.1.2 Multi-mode fiber (based on number of modes)

The light ray flowing through multimode fiber can take on a variety of modes. The cladding has a diameter of (40 μm) and the core has a diameter of (70 μm). The difference in relative refractive index is also bigger than in single mode fiber. Due to

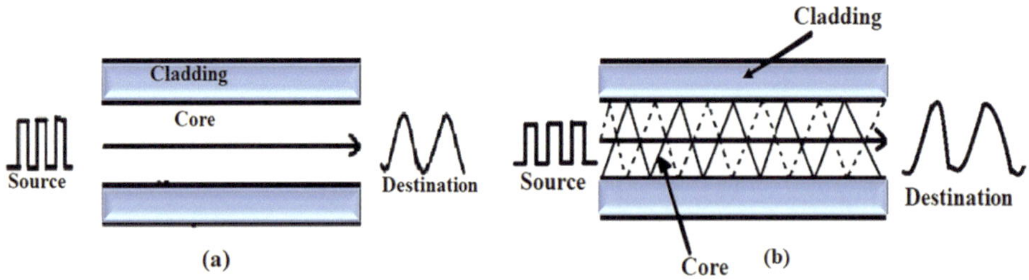

Figure 2.
(a) Single mode optical fiber; and (b) multi mode optical fiber.

multimode dispersion, signal quality suffers. Due to signal dispersion and attenuation, it is not ideal for long-distance transmission [42] (shown in **Figure 2b**).

1.1.1.3 Step index fiber (based on refractive index)

The refractive index of step-index fibers is discontinuous at the interface between the cladding and the core, and the core is constant. Light rays pass through it as meridional rays that cross the fiber axis at the core-cladding interface on every reflection as shown in **Figure 3a**.

1.1.1.4 Graded index fiber (based on refractive index)

The core of this fiber has a non-uniform refractive index that declines gradually from the centre to the core-cladding interface. The refractive index of the cladding is uniform. Light rays in the form of skew rays or helical rays propagate through it as shown in **Figure 3b**. At no point does it cross the fiber axis. The refractive index of the core in a gradient index fiber is higher in the centre and gradually drops as the interface approaches [43, 44].

1.1.2 Design OF a basic fiber optic gas sensing system

A gas sensor is a device that measures target gas molecules present in a given environment. When gas molecules encounter the sensor's solid receptors, a potential difference occurs, that is converted into an electrical signal. The sensor's gas sensitivity and the selectivity are dependent upon the reaction of the sensing materials with the gases [45]. The sensing materials basically used for all types of gas sensors include polymers, organic monolayers, ceramics, semiconductors, porous nanomaterial and nanostructured materials (nanorods, nanotubes, nanodots). Optical fiber sensors have been projected as a good choice, to considering their benefits compared to other available sensing technologies, as they are immune to electromagnetic interferences [46], do not require any electric power to work, possibility of multiplexation, ability to perform in remote areas and harsh environmental conditions [27]. A block diagram of a common fiber optic gas sensor is shown in **Figure 4**. A light source, a signal input optical fiber, a signal output optical fiber, and a detector make up a fiber-optic gas sensing system (optionally the system may include other components such as an optical modulator and a demodulator). The essential idea is that the light from the source is transferred to the sensor element via the incident optical fiber.

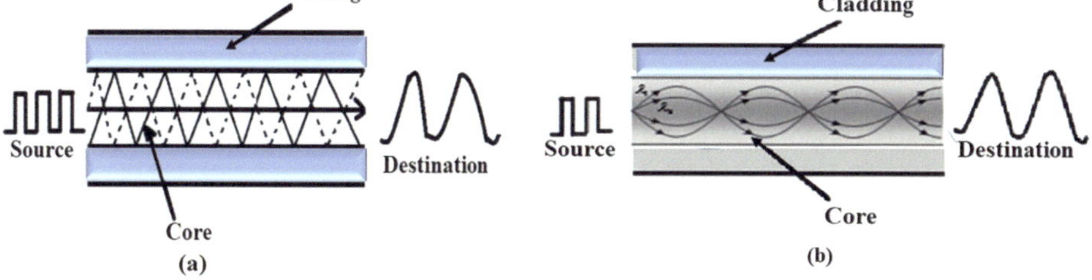

Figure 3.
(a) Step index optical fiber; and (b) graded index optical fiber.

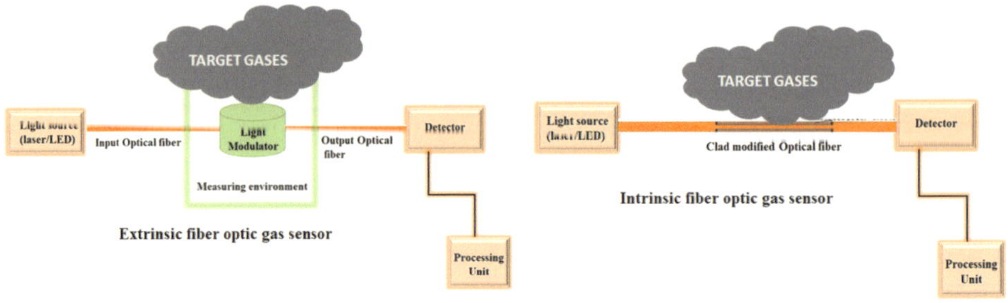

Figure 4.
Block diagram of extrinsic and intrinsic fiber optic gas sensor.

The measured parameters are obtained by modulating the optical properties of light, such as intensity, wavelength, frequency, phase, and polarisation state, in the sensor element and sending them to the optical detector through the outgoing fiber. Intrinsic and extrinsic optical fiber sensors are two types of sensors. The intrinsic optical fiber sensor relies on the optical fiber's own sensitivity to environmental variables. In an intrinsic sensor, specific regions in the fiber cable perform sensing function, thereby detector is not exposed to the light source. However, in extrinsic sensors, the fiber optic cable is used as a data transmission line, wherein the optical cable carries light from an optical source externally to the sensing region. Most of the fiber optic sensors used in VOC gas detection are intrinsic sensors.

1.1.3 Classification of fiber optic gas sensing (FOS) system

The fiber optic sensing region is primarily divided into four kinds of approaches namely the Fabry-Perot Interferometer (FPI), Michelson Interferometer (MI), Mach-Zehnder Interferometer (MZI), Sagnac Interferometer (SI), based on interferometric techniques [47, 48] as shown in **Figure 5**. These approaches are known for

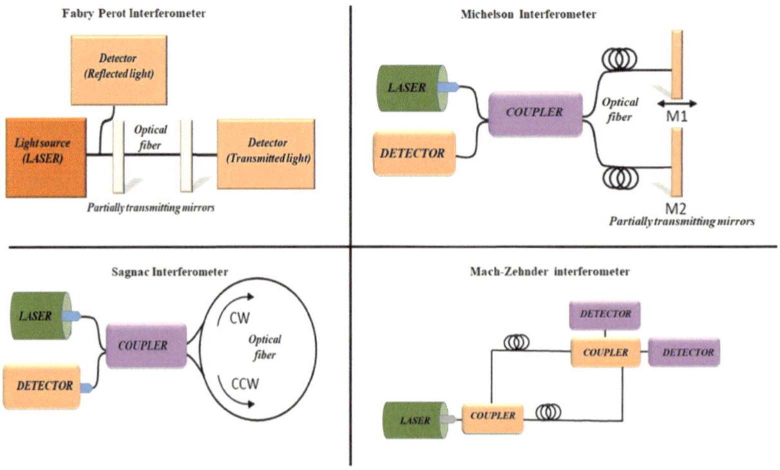

Figure 5.
Fiber optic sensors based on interferometric techniques.

their diverse geometries, operating principles, and their sensitivity. The Michelson, Mach-Zehnder, and Sagnac interferometers use a two-beam interference principle, whereas the Fabry-Perot interferometer uses a multi-beam interference principle. An interferometer's basic operation is to split an incident light beam into two portions, the reference beam and the sensing beam, the latter of which is changed by a variable of interest that we want to measure. The two beams are then recombined to form an interference pattern that allows information about the desired variable to be recovered. Other fiber-optic geometries based on the SPR principle, gratings, and the evanescent wave absorption (EWA) principle exist. An overview of various kinds of Fiber optic gas sensors is displayed in **Figure 6**.

1.1.3.1 Fabry-Perot interferometer (FPI)

Because of its simplicity, sensitivity, and cost-effectiveness, this is the most commonly used interferometric geometry for chemical sensing. The FPI is made up of two parallel reflecting mirrors that are spaced apart. It's known as an etalon. A single light beam enters the FP cavity and is multiplicatively internally reflected. The superposition of both reflected and transmitted beams at the two parallel surfaces causes interference. The FP cavity can be created inside or at the end of an optical fiber in the case of FOS. Extrinsic FPI and intrinsic FPI are the two types of FPI that exist. An extrinsic FPI uses the reflections from the cavity surfaces to create a cavity at the end of the optical cable. As a result, an extrinsic FPI is defined as a cavity material that is not made up entirely of fiber. The cavity and reflections arise within the fiber itself in the case of an intrinsic FPI [49]. The FPI is easy to build and responds well to external inputs.

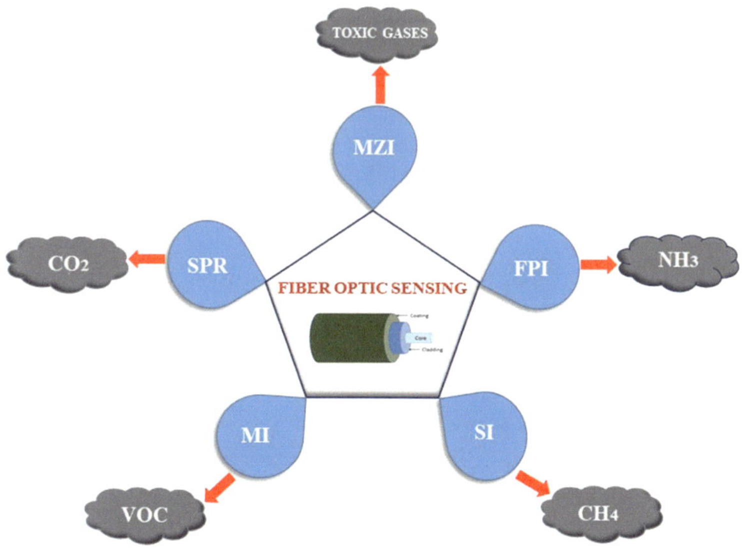

Figure 6.
An overview of various kinds of fiber optic gas sensors.

1.1.3.2 The Michelson interferometer (MI)

The Michelson interferometer (MI) is one of the most frequent and straightforward interferometers. The beams are separated and recombined using a single beam splitter (coupler). The reference beam is reflected by a fixed mirror in a traditional Michelson interferometer, while the sensing beam is reflected by a moving mirror [50].

1.1.3.3 The Mach-Zehnder interferometer (MZI)

The light beam is split into two pieces that go through the reference and detecting arms, respectively, before being recombined in this arrangement. The optical channels are only crossed once, unlike in the Michelson interferometer. Two photodetectors are installed in each of the interferometer's outputs. The phase shift caused by variations in the sensing arm caused by temperature, strain, or refractive index changes, among other things, can be quantified using the intensities detected by both detectors. This interferometer can be implemented in integrated optics, which protects the device, reduces its size, and allows it to be used for other purposes like biosensing [51, 52].

1.1.3.4 The Sagnac interferometer (SI)

The input beam is split into two equal-intensity counter-propagating beams that move along a ring route. The Sagnac effect occurs when the Sagnac interferometer is spun, causing a relative phase change between the beams. The interference spectrum is determined by the setup's angular frequency. Currents, acoustic waves, strain, and temperature can all be monitored using this interferometer [53].

1.1.3.5 Surface plasmon resonance (SPR)

Because of its light guidance inside fiber based on the TIR effect, simple and flexible design, compactness, and remote sensing ability, optical fiber has proven to be a very useful instrument in the SPR sensing system. When light travels through the core of the fiber, some of it will pass through the cladding region. The plasmonic metal surface interacts with the evanescently leaked light, which stimulates the surface's free electrons. Surface plasmon is created when the evanescent field and free electron

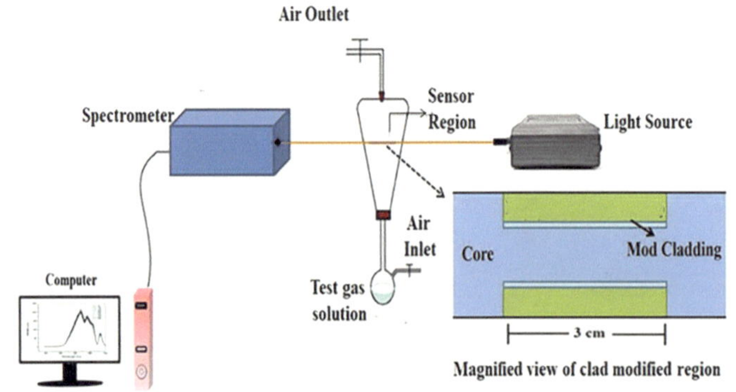

Figure 7.
Fiber optic gas sensing set-up using evanescent wave absorption based on experiment [55].

surface frequencies are in resonance. This generated wave propagates along the metal-dielectric interface and exhibits both charge motion in the metal (surface plasmon) and electromagnetic wave in the air or dielectric (polariton) [54].

1.1.3.6 Evanescent wave absorption type (EWA)

The leakage/loss of electromagnetic energy at the interface of the core and the cladding medium during a total internal reflection (TIR) event determines the evanescent wave absorption based sensing phenomenon. If the diameters of a fiber are big in comparison to the wavelength of light, this method may be applied. When a propagating beam's incident angle (θi) is greater than the critical angle (θc), TIR causes the ray to reflect back from the core-cladding contact. A small fraction of the guided wave's energy enters the cladding medium at each TIR event, generating an electromagnetic field known as a "evanescent wave." The evanescent wave is greatly affected by variations in refractive index at the core-cladding contact, which provides a foundation for sensors based on evanescent wave absorption as shown in **Figure** 7 [55, 56].

2. Materials used in gas sensing

2.1 Spin frustrated Multiferroics: A novel frustrated ordering studied in gas sensors

Spin frustration in ME multiferroics have received enormous attention ever since the concept was introduced in 1977 by Gerard Toulouse. It is an important feature in the field of magnetism and multiferroics, as it stems from the relative arrangement of spins. Interestingly, it is not the strength of ME coupling or the high polarization that makes these materials unique; in fact, such systems have weak coupling and low magnitude of polarization compared to other multiferroics [57–60]. Still, the reason for its long sought control of electric properties by magnetic fields, lies in the magnetic origin of their ferroelectricity, which is induced by the presence of complex spin structures, characteristic of frustrated magnets [61]. Basically, in most of the magnetic solids, the magnetic moments form a long range ordered ferromagnetic (↑↑ moments) or antiferromagnetic (↑↓ moments) structure when cooled below a certain temperature. However, in "frustrated" magnetic solids, there is no long range ordering at low temperature phase. Instead, they develop high degenerate states with short range spin correlations. Typical examples of geometrically frustrated structures are 2D triangular or tetrahedron non collinear magnetic structures, where the lattice consists of triangular or tetrahedral arrangements of anti-ferromagnetically (AFM) coupled

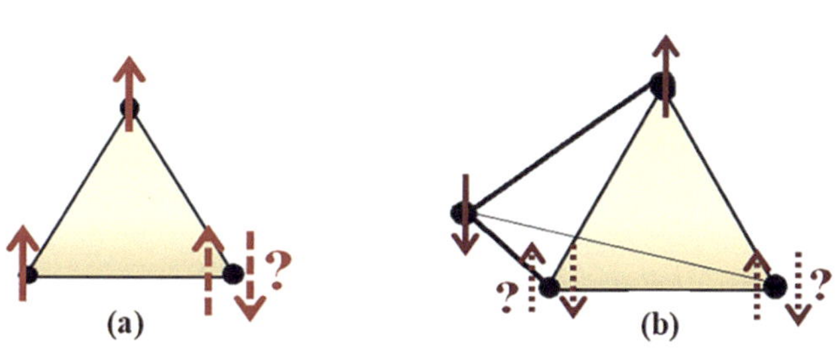

Figure 8.
Spin frustration in 2D (a) triangular; and (b) tetrahedron magnetic structure.

spins as shown in **Figure 8**. The spin on the third site can be in any one direction, either up (or) down, but then the interaction between this site and the other two sites will be different. As a consequence, the third site will move closer to one site and away from the other, which will break the symmetry and induce ferroelectricity. This kind of lattice geometry creates large degeneracy of ground states within which the system can fluctuate with almost no energy expenditure, even below few milli Kelvin temperatures. A one dimensional anti-ferromagnetic material has ground state in alternating series of spins: up, down, up, down, etc. But, in 2D equilateral triangular AFM lattice, multiple ground states can occur, with three spins, one on each vertex. If each spin can take on only two values (up or down), there are $2^3 = 8$ possible states of the system, six of which are ground states. Two situations, which does not favor ground states are when all three spins are up or are all down. However, in other six states, there will be two favorable interactions and one unfavorable one. This illustrates spin frustration: the inability of the magnetic system to find a single ground state [58, 60].

2.1.1 Factors to consider for sensor interrogation of spin frustrated multiferroic materials

The main factors that are critical in gas detection, particularly when employing multiferroic materials as sensing materials are

1. Porosity, Surface area and Crystallinity of the material

2. Sensitivity and Selectivity of the material

3. Dynamic Response and Stability of the material

4. Doping or Substitution effect in the material

5. Temperature and Humidity of the surrounding environment

2.2 Few spin frustrated Multiferroics

This section discusses a few spin frustrated multiferroics that have been studied as gas sensors.

2.2.1 $Bi_2Fe_4O_9$

The Spin frustrated Mullite type $Bi_2Fe_4O_9$ (BFO), yielded as an undesired secondary phase from the production of $BiFeO_3$ [61], is being extensively evaluated due to its range of industrial applications in the form of catalyst for oxidation of ammonia, sensors for magnetic field detection, and even as memory storage devices [62–66]. However, the most common issue experienced in this single-phase multiferroic system is the occurrence of weak magnetic/electric fields at room temperature. The Fe6 atoms that are synchronized in an octahedral fashion, interact ferromagnetically with one another while the Fe4 atoms synchronized tetrahedrally towards the inner axis of the unit cell interact in antiferromagnetic order thus ensuring a spin frustrated configuration with weak magnetic order beyond 264 K and weak ferroelectric ordering below 250 K [67]. As documented in reports, partially doping any transition metal ion or rare-earth (RE) ion into either of the sites of BFO can enhance the gas sensing parameters such as its conductivity,

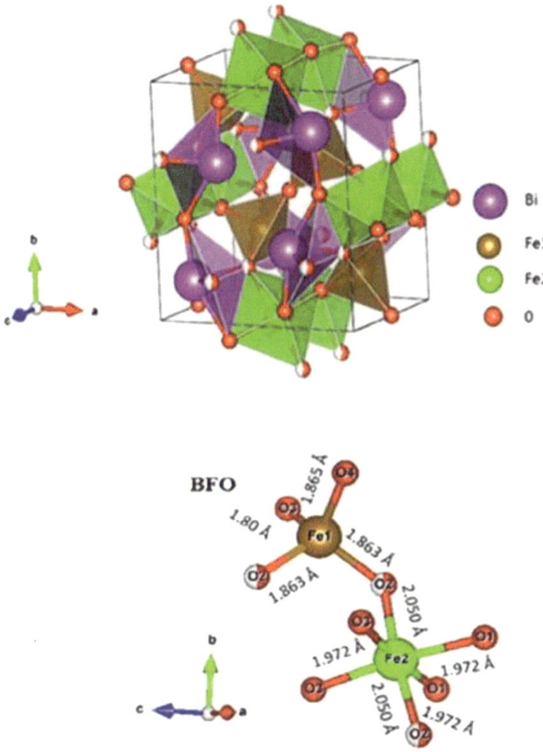

Figure 9.
Crystal structure of $Bi_2Fe_4O_9$ and the bond angles and bond lengths of the material [71].

catalytic activity required for oxygen adsorption, electron mobility, chemical stability, and enhanced sensitivity [68–71]. Given the usage of a magnetic dopant, it is important to understand the nature of magnetism in this sample $Bi_2Fe_4O_9$ samples exhibited significant changes in magnetic and electrical characteristics, which are necessary for improved photocatalytic and gas sensing applications [68, 70]. $Bi_2Fe_4O_9$ crystal structure consists of two Fe sites, one consisting of the Fe in the tetrahedral coordination (Fe1) and other Fe in octahedral coordination (Fe2) as shown in **Figure 9**. $Bi_2Fe_4O_9$ is widely used for the detection of VOC test gases at laboratory scale owing to their good selectivity, good long-term stability, low synthesis cost and most importantly its ability to sense at room temperature. Recently Subha et al. [55] reported rare earth Nd doped $Bi_2Fe_4O_9$ (BNFO3) as a gas sensing material using Fiber optic evanescent wave absorption set up to detect VOC's such as ammonia, ethanol, methanol and acetone at room temperature as shown in **Figure** 7. The improved sensing ability in the Nd-doped $Bi_2Fe_4O_9$ sample was seen towards ammonia vapours as shown in **Figure 10** with an extremely short response and recovery time of 40 sec and 48 sec, respectively, thereby making them an efficient fiber optic ammonia gas sensor. The same group also investigated room temperature gas sensing ability in Gd doped $Bi_2Fe_4O_9$ sample, and observed enhanced sensing capability towards ethanol gas at 500 ppm. The clad modified fiber optic gas sensing studies suggested that Gd doping improved ethanol gas sensing ability from 0 to 500 ppm at room temperature compared to host $Bi_2Fe_4O_9$ with high sensitivity, quick reaction and recovery period of around 38 s and 67 s, respectively. All of these findings highlight the $Bi_2Fe_4O_9$ with enhanced gas sensing capability suitable for industrial applications [26].

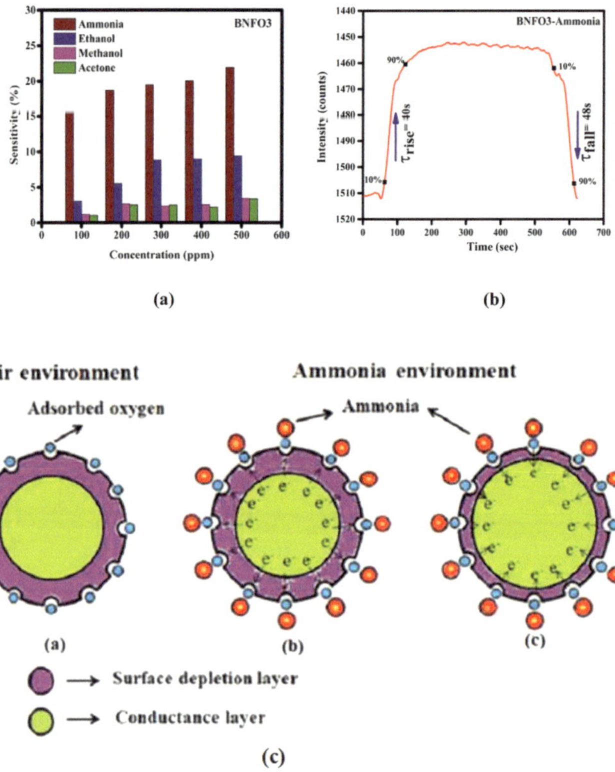

Figure 10.
(a) Sensitivity of BNFO3 towards various VOC gas vapors; (b) response analysis of BNNFO3 sample; and (c) gas sensing mechanism in BNFO3 sample [55].

2.2.2 $YMnO_3$

The usage of YMnO3 perovskite materials in gas sensing properties has recently garnered attention. Such system shows high reactive and stable behaviour over multiple oxidation/reduction states in the experimental cycles. Yttrium and rare-earth manganites in $RMnO_3$ oxide stabilises in two structural phases. The hexagonal phase (space group P63cm, Z = 6) forms with R = Ho, Er, Tm, Yb, Lu, or Y, which have a small ionic radius whereas the orthorhombic phase (space group Pnma, Z = 4) forms with R = La, Ce, Pr, Nd, Sm, Eu, Gd, Tb, or Dy, which possess a larger ionic radius. The orthorhombic phase exhibits ferromagnetic ordering, whereas the hexagonal phase shows both ferromagnetic and ferroelectric ordering [72–74]. Hexagonal $YMnO_3$ oxide, with an orthorhombic crystal structure (**Figure 11**) is an excellent material for non-volatile memory and metal-ferroelectric-semiconductor (MFS) devices, because of their coupled magnetic and ferroelectric behaviour [76–79]. Such electromagnetic multiferroics, which exhibit simultaneous ferroelectricity and magnetism, can be exploited in both electrical and magnetic applications. To date, many reports have focused on $YMnO_3$ based materials because of their unique dielectric, ferroelectric, and magnetic properties [75, 80–83].

Recently, gas sensing properties of $YMnO_3$ was tested on LPG, H_2, CO and H_2S gases and among the gases it showed maximum sensitivity to H_2S gas when compared to other gases. The gas sensor fabrication is shown in **Figure 12a** wherein spin frustrated

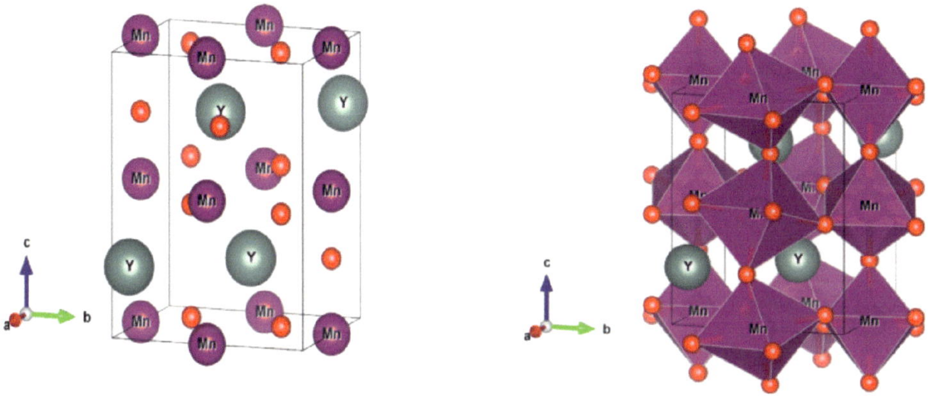

Figure 11.
The orthorhombic crystal structure of $YMnO_3$, showing the magnetic element Mn [75].

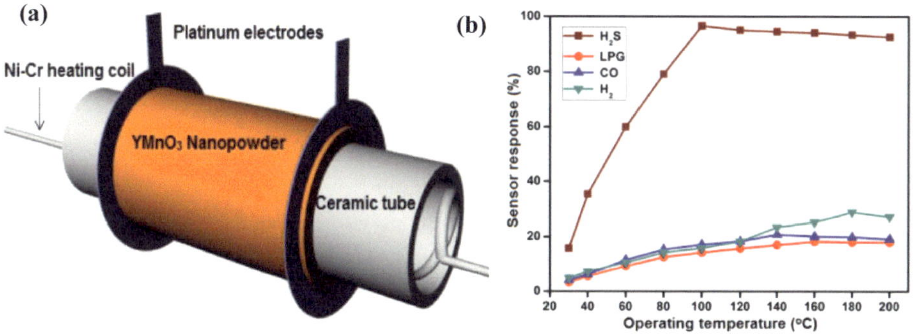

Figure 12.
(a) Schematic diagram of $YMnO_3$ sensor element; and (b) sensor response towards 500 ppm of reducing gases as a function of operating temperature [78].

$YMnO_3$ of 50 μm thickness is coated inside a cylindrical tube of 8 mm length and 2 mm diameter. Nichrome wire was fixed inside the tube to heat the chamber and to carry out the temperature dependent sensitivity measurement. Chromel alumel thermocouple was used for monitoring the temperature of the tube [78]. Voltage drop in with the application of 10 V was used for sensing the gas. The sensing property of $YMnO_3$ was measured in the presence gases such as CO, H_2, H_2S and LPG at a concentration of 500 ppm. **Figure 12b** shows the dynamic sensing properties as a function of operating temperature for the $YMnO_3$ different gases. The response of the sensor increases with temperature up to 100°C, and then it decreases with temperature. The $YMnO_3$ sensor element shows a 96% response for an operating temperature of 100°C for H_2S gas. Other reducing gases such as CO, H_2, and LPG show comparatively low responses [78].

2.2.3 $Ni_3V_2O_8$

$Ni_3V_2O_8$ is an important system of spin frustrated magnet at low temperatures it undergoes a series of competing magnetic ordering. $Ni_3V_2O_8$ has orthorhombic crystal structure with space group Cmca, which is characterized by triangular lattices and short range antiferromagnetic interactions [84]. In $Ni_3V_2O_8$, the magnetic

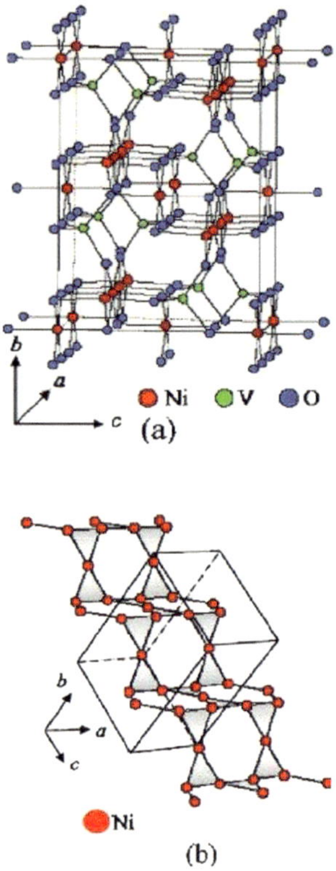

Figure 13.
(a) Shows the crystal structure of $Ni_3V_2O_8$; and (b) Kagoma stair case showing only the Ni atoms.

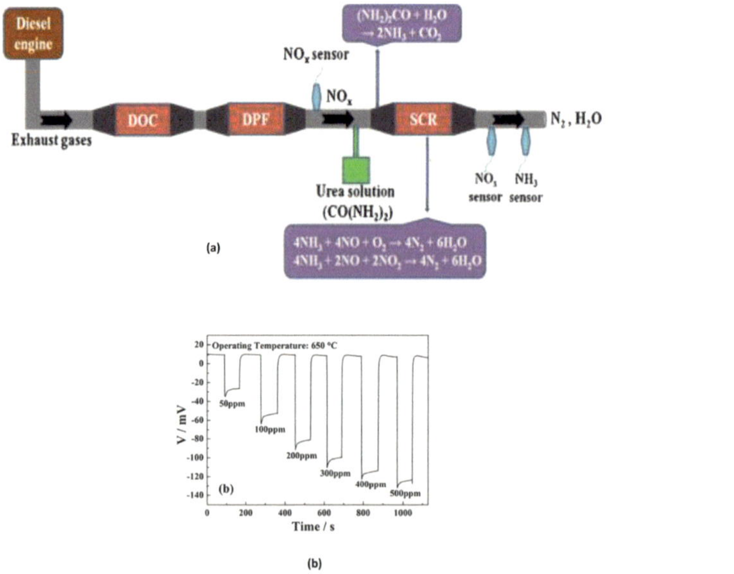

Figure 14.
(a) Schematic view of exhaust gases after treatment system; and (b) response transients curve for NH_3 in the range of 50–500 ppm at 650°C [87].

lattice is made up of magnetic ion Ni^{2+} (S = 1, d8), based on an anisotropic Kagomé lattice. Non-magnetic VO_4 tetrahedra layer separates the magnetic layers. Due to this separation the geometric frustration is reduced and this allows long-range magnetic ordering in the material. Ni^{2+} has two kind of positions that is "spine" and "cross tie" correspondingly. The deviation from the ideal Kagomé geometry introduces many new interactions that relieve the frustration of underlying Kagomé antiferromagnet interacting. The structure of $Ni_3V_2O_8$ is shown in **Figure 13**.

Recently, $Ni_3V_2O_8$ is used as important materials for gas sensing application [85–87]. $Ni_3V_2O_8$ showed the high sensitivity to Ammonia gas. To control the amount NH_3 that is sent to the air, a powerful closed loop control system based on the NH_3 is required and the schematic of the automobile based sensors are given in **Figure 14a**. The sensor showed the optimal behaviour at 650°C beyond this temperature the behaviour decreases (**Figure 14b**). The sensor 90% response and recovery times of the sensor to 500 ppm NH_3 were approximately 2 and 10 s, respectively [87].

3. Applications and challenging aspects of fiber optic sensors

Fiber optic sensors (FOS) have been employed in a variety of harsh environmental applications where other sensor types typically fail, due to its unique characteristics such as electromagnetic immunity, intrinsic safety, chemical and heat resistance, and exceptionally small size. Some innovative FOS solutions have been used in high-temperature, high-pressure, and possibly explosive environments, such as oil and gas wells [88], pipelines [89], and in turbines testing and engine testing [90–92]. It is well recognized that fiber optic sensors play a vital role in many applications. Fiber optic sensors are expected to be utilised to improve the efficacy and cost-effectiveness of many electronic goods, given the vast variety of benefits that fiber optic sensing offers in multiple industries. The usage of fiber optic sensors in environmental monitoring is quadrupling, which is critical for guaranteeing adequate food and water supplies, identifying potential airborne contaminants, and safeguarding structures from corrosion. Water safety, agriculture, transportation, smart structure protection, and biomedical monitoring are all projected to see an increase in environmental monitoring. Further, the future prospects of fiber optic sensors towards various technological aspects are represented in **Figure 15**,

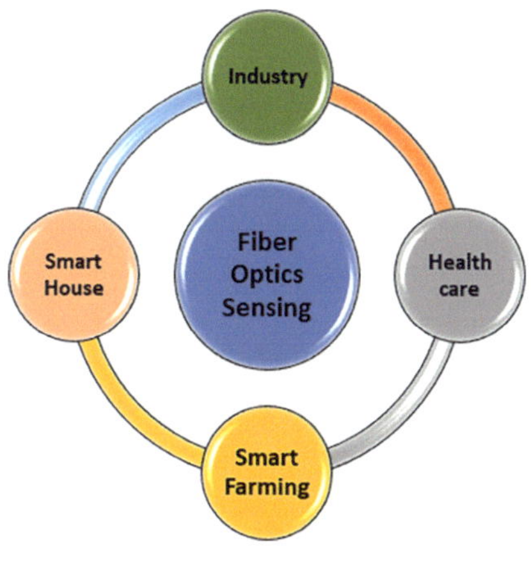

Figure 15.
Future prospects of fiber optic sensors towards various technological aspects.

wherein this technology would be employed in smart city initiatives, created specifically for tough and challenging atmosphere. Despite its interesting solutions in environmental monitoring, Clinical environment is still facing challenges in some ways, particularly if FOS is poorly constructed. In a surgical room with a critically ill patient, for example, the instrumentation must be as simple as possible so that the medical personnel may focus on the emergency rather than how the sensor should be attached or the system configured. With a "plug and play" philosophy, it should be as simple as it is with other existing electrical devices. Most practitioners do not yet have the background associated with FOS technologies, and they should not only gain complete confidence in this new technology that is slowly gaining traction in their environment, but also be completely at ease with the potential addition of new steps to their daily medical procedures. In practise, it signifies that an efficacious incorporation of FOS technology is highly reliant on a thorough understanding of medical applications and related clinical measures so as to benefit more from the optical sensing technology.

Despite the multiple challenges and impediments to sensor deployment in smart technology, R&D breakthroughs would result in the widespread availability of low-cost and precise sensors for monitoring water, soil nutrition, temperature, and humidity. Fiber optic sensor technology will continue to grow slowly and steadily over the next few years. Researchers in the field of photonics will continue to be fascinated by fiber optic sensors. They're looking forward to developing new technologies and seeing what these sensors can do for the sensing and instrumentation industries. We recognise that this study may not be exhaustive in all categories, but it is an attempt to provide readers with an overview and the most straightforward way when developing and researching Fiber optic gas sensors.

Author details

Subha Krishna Rao[1*], Rajesh Kumar Rajagopal[2] and Gopalakrishnan Chandrasekaran[3*]

1 Centre for Materials Engineering and Regenerative Medicine, Bharath Institute of Higher Education and Research, Chennai, India

2 School of Nanoscience and Technology, IIT Kharagpur, Kharagpur, India

3 Department of Physics and Nanotechnology, SRM IST, Chennai, Tamil Nadu, India

*Address all correspondence to: subhha2010@gmail.com and cgknano@gmail.com

References

[1] Kapse VD. Preparation of Nanocrystalline spinel-type oxide materials for gas sensing applications. Research Journal of Chemical Sciences. 2015;**5**:7-12

[2] Sberveglieri G, editor. Gas Sensors: Principles, Operation and Developments. Berlin, Germany: Springer Science & Business Media; 2012

[3] Symons EA. Catalytic gas sensors. In: Gas Sensors. Dordrecht: Springer; 1992. pp. 169-185

[4] Bakker E, Telting-Diaz M. Electrochemical sensors. Analytical Chemistry. 2002;**74**:2781-2800. DOI: 10.1021/ac0202278

[5] Hodgkinson J, Tatam RP. Optical gas sensing: A review. Measurement Science and Technology. 2013;**24**:012004. DOI: 10.1088/0957-0233/24/1/012004

[6] Bhattacharyya P. Technological journey towards reliable microheater development for MEMS gas sensors: A review. IEEE Transactions on Device and Materials Reliability. 2014;**14**:589-599

[7] Cheeke JDN, Wang Z. Acoustic wave gas sensors. Sensors and Actuators B: Chemical. 1999;**59**:146-153

[8] Akubik WP. Surface acoustic wave-based gas sensors. Thin Solid Films. 2011;**520**:986-993

[9] Joe HE, Yun H, Jo SH, Jun MBG, Min BK. A review on optical fiber sensors for environmental monitoring. International Journal of Precision Engineering and Manufacturing—Green Technology. 2018;**5**:173-191

[10] Pawar D, Kale SN. A review on nanomaterial-modified optical fiber sensors for gases, vapors and ions. Microchimica Acta. 2019;**186**:1-34

[11] Available from: https://www.marketsandmarkets.com/Market-Reports/optical-sensing-market-197592599.html

[12] Bontempelli G, Comisso N, Toniolo R, Schiavon G. Electroanalytical sensors for nonconducting media based on electrodes supported on perfluorinated ion-exchange membranes. Electroanalysis. 1997;**9**:433-443

[13] Yamazoe N, Sakai G, Shimanoe K. Oxide semiconductor gas sensors. Catalysis Surveys from Asia. 2003;**7**:63-75

[14] Emelin EV, Nikolaev IN. Sensitivity of MOS sensors to hydrogen, hydrogen sulfide, and nitrogen dioxide in different gas atmospheres. Measurement Techniques. 2006;**49**:524-528

[15] Kavitha G, Kumar JV, Arulmozhi R, Kamath SM, Priya AK, Rao KS, et al. 2D graphene supported nickel oxide nano-composite for fiber optic ethanol gas sensing, removal of azo dye, and biological activity. Journal of Materials Science: Materials in Electronics. 2022;**33**:9498-9511

[16] Arshak K, Moore E, Lyons GM, Harris J, Clifford S. A review of gas sensors employed in electronic nose applications. Sensor Review. 2004;**24**:181-198

[17] Munoz R, Sivret EC, Parcsi G, Lebrero R, Wang X, Suffet IH, et al. Monitoring techniques for odour abatement assessment. Water Research. 2010;**44**:5129-5149

[18] Stetter JR, Penrose WR. Understanding chemical sensors and

chemical sensor arrays (electronic noses):Past, present, and future. Sensors Update. 2002;**10**:189-229

[19] Wilson AD. Review of electronic-nose technologies and algorithms to detect hazardous Chemicals in the Environment. Procedia Technology. 2012;**1**:453-463

[20] Boeker P. On "Electronic Nose" methodology. Sensors and Actuator B Chemical. 2014;**204**:2-17

[21] Brzózka Z, Wróblewski W. Sensory Chemiczne. Warsaw, Poland: Oficyna Wydawnicza Politechniki Warszawskiej; 1998

[22] Wilson AD, Baietto M. Applications and advances in electronic-nose technologies. Sensors. 2009;**9**:5099-5148

[23] Kalai Priya A, Yogesh GK, Subha K, Kalyanavalli V, Sastikumar D. Synthesis of silver nano-butterfly park by using laser ablation of aqueous salt for gas sensing application. Applied Physics A: Materials Science and Processing. 2021;**127**:292-304

[24] Renganathan B, Rao SK, Ganesan AR, Deepak A. High proficient sensing response in clad modified ceria doped tin oxide fiber optic toxic gas sensor application. Sensors and Actuators A: Physical. 2021:**332**:113114

[25] Yadav PK, Srivastava R, Chaurasiya N. Optical fiber sensor : Review and applications. IJCRT. 2018;**6**(1):0-5

[26] Rao SK, Priya AK, Kamath SM, Abhinav EM, Renganathan B, Jeyadheepan K, et al. Unraveling the potential of Gd doping on mullite BiFeO for fiber optic ethanol gas detection at room temperature. Materials Chemistry and Physics. 2021;**278**:125646

[27] Pawar D, Kale SN, Birefringence manipulation in tapered polarization-maintaining photonic crystal fiber Mach-Zehnder in- terferometer for refractive index sensing. Sens Actuators A Phys, 2016;**252**:180-184

[28] Vitoria I, Zamarreño CR, Ozcariz A, Matias IR. Fiber optic gas sensors based on lossy mode resonances and sensing materials used therefor: A comprehensive review. Sensors. 2021;**21**:11-26

[29] Roriz P, Ramos A, Santos JL, Simões, JA. Fiber Optic Intensity Modulated Sensors: A Review in Biomechanics. Photonic Sens. 2012;**2**(4):315-330

[30] Hull D, Souders J. Principles of Fiber Optic Communication. Waco, Texas: CORD Communications; 2008

[31] Agarwal T. What are the basic elements of a fiber optic communication system? The Budding Electronic Engineer's Knowledge Space. 2010:1-4

[32] Patrick HJ, Williams GM, Kersey AD, Pedrazzani JR, Vengsarkar AM. Hybrid Fiber Bragg Grating/long Period Fiber Grating Sensor for Strain/temperature Discrimination. IEEE Photonics Technology Letters. 1996;**8**(9):1223-1225

[33] Bao Y, Huang Y, Hoehler MS, Chen G. Review of fiber optic sensors for structural fire engineering. Sensors (Switzerland). 2019;**19**:1-23

[34] Mahata P, Mondal SK, Singha DK, Majee P. Luminescent rare-earth-based MOFs as optical sensors. Dalton Transactions. 2017;**46**:301-328

[35] Zhang Y-N, Zhao Y, Zhou T, Wu Q. Applications and developments of on-chip biochemical sensors based on optofluidic photonic crystal cavities. Lab on a Chip. 2018;**18**:57-74

[36] Li J, Yim D, Jang W-D, Yoon J. Recent progress in the design and applications of fluorescence probes containing crown ethers. Chemical Society Reviews. 2017;**46**:2437-2458

[37] Wu D, Sedgwick AC, Gunnlaugsson T, Akkaya EU, Yoon J, James TD. Fluorescent chemosensors: The past, present and future. Chemical Society Reviews. 2017;**46**:7105-7123

[38] Wu D, Chen L, Lee W, Ko G, Yin J, Yoon J. Recent progress in the development of organic dye based near-infrared fluorescence probes for metal ions. Coordination Chemistry Reviews. 2018;**354**:74-97

[39] Lopez O, Haboucha A, Chanteau B, Chardonnet C, Amy-Klein A, Santarelli G. Ultra-stable long distance optical frequency distribution using the internet fiber network. Optics Express. 2012;**20**(21):23518-23526

[40] Addanki S, Amiri IS, Yupapin P. Review of optical fibers-introduction and applications in fiber lasers. Results in Physics. 2018;**10**:743-750

[41] Di Sante R. Fiber optic sensors for structural health monitoring of aircraft composite structures: Recent advances and applications. Sensors. 2015;**15**:18666-18713

[42] Personick S. Review of fundamentals of optical fiber systems. IEEE Journal on Selected Areas in Communications. 1983;**1**:373-380

[43] Kouhdaragh V, Tarchi D, Vanelli-Coralli A, Corazza GE. Smart Grid Inspired Future Technologies. Vol. 24. Liverpool, UK: Springer Champ; 2017:**16**

[44] Zhou X, Chen Z, Zhou H, Hou J. Mode-field adaptor between large-mode-area fiber and single-mode fiber based on fiber tapering and thermally expanded core technique. Applied Optics. 2014;**53**:5053-5057

[45] Padvi MN, Moholkar AV, Prasad SR, Prasad NR. A critical review on design and development of gas sensing materials. Engineering and Science. 2021;**15**:20-37

[46] Joe HE, Yun H, Jo SH, Jun MBG, Min BK. A review on optical fiber sensors for enironmental monitoring. International Journal of Precision Engineering and Manufacturing—Green Technology. 2018;**5**:173-191

[47] Ran Z, He X, Rao Y, Sun D, Qin X, Zeng D, et al. Fiber-optic microstructure sensors: A review. Photonic Sensors. 2021;**11**:227-261

[48] Lee BH, Kim YH, Park KS, Eom JB, Kim MJ, Rho BS, et al. Interferometric fiber optic sensors. Sensors. 2012;**12**:2467-2486

[49] Islam MR, Ali MM, Lai MH, Lim KS, Ahmad H. Chronology of Fabry-Perot interferometer fiber-optic sensors and their applications: A review. Sensors. 2014;**14**:7451-7488

[50] Liu D, Xie Y, Xin G, Li Z. Fiber-optic michelson interferometer sensor fabricated by femtosecond lasers. Sensors & Transducers. 2013;**160**:215-219

[51] Li L, Xia L, Xie Z, Liu D. All-fiber Mach-Zehnder inter-ferometers for sensing applications. Optics Express. 2012;**20**:11109-11120

[52] Pawar D, Kale SN. Birefringence manipulation in tapered polarization-maintaining photonic crystal fiber Mach-Zehnder in-terferometer for refractive index sensing. Sensors and Actuators A: Physical. 2016;**252**:180-184

[53] Pospíšilová M, Kuncová G, Trogl J. Fiber-Optic chemical sensors and fiber-optic bio-sensors. Sensors; 2015;**15**:25208-25259

[54] Srivastava SK, Verma R, Gupta BD. Surface plasmon resonance based fiber optic sensor for the detection of low water content in ethanol. Sensors Actuators, B Chemical. 2011;**153**:194-198

[55] Krishnarao S, Priya K, Manjunath K. Unequivocal evidence of enhanced room temperature sensing properties of clad modified Nd doped mullite $Bi_2Fe_4O_9$ in fiber optic gas sensor. Journal of Alloys and Compounds. 2020;**838**:155603

[56] Sharma AK, Gupta J, Sharma I. Fiber optic evanescent wave absorption-based sensors: A detailed review of advancements in the last decade (2007-18). Optik (Stuttg). 2019;**183**:1008-1025

[57] Moessner R, Ramirez AP. Geometrical frustration. Physics Today. 2006;**59**:24

[58] Ramirez AP. Strongly geometrically frustrated magnets. Annual Review of Materials Science. 1994;**24**:453

[59] Harris M, Zinkin M. Frustration in the pyrochlore antiferromagnets. Modern Physics Letters B. 1996;**10**:417

[60] Greedan JE. Geometrically frustrated magnetic materials. Journal of Materials Chemistry. 2001;**11**:37

[61] Cheong S-W, Mostovoy M. Multiferroics: A magnetic twist for ferroelectricity. Nature Materials. 2007;**6**:13-20

[62] Raghavan CM, Kim JW, Kim JW, Kim SS. Structural, electrical and multiferroic properties of La-doped mullite $Bi_2Fe_4O_9$ thin films. Materials Research Bulletin. 2015;**70**:279-283

[63] Zakharchenko NI. Catalytic properties of the system Fe 2 O 3—Bi 2 O 3 in ammonia oxidation. Kinetics and Catalysis. 2001;**42**:747-753

[64] Eerenstein W, Mathur ND, Scott JF. Multiferroic and magnetoelectric materials. Nature. 2006;**442**:759-765

[65] Martin LW, Chu YH, Ramesh R. Advances in the growth and characterization of magnetic, ferroelectric, and multiferroic oxide thin films. Materials Science and Engineering R: Reports. 2010;**68**:89-133

[66] Meher Abhinav E, Jaison DA, Sundararaj SK, Rao, et al., A thermomagnetic analysis of an eco-friendly nano-sized dysprosia for energy efficient cryo-cooling systems. Journal of Nanoparticle Research. 2021;**23**:1-13

[67] Singh AK, Kaushik SD, Kumar B, Mishra PK, Venimadhav A, Siruguri V, et al. Substantial magnetoelectric coupling near room temperature in Bi2 Fe_4 O_9. Applied Physics Letters. 2008;**92**:2006-2009

[68] Rao SK, Althaf R, Chandrasekaran G. Investigation of room temperature multi-functional properties of Nd doped Mullite $Bi_2Fe_4O_9$. Vacuum. 2019;**172**:109109

[69] Wu S, Li C, Wei W, Wang H, Song Y, Zhu Y, et al. Nd-doped SnO_2: Characterization and its gas sensing property. Journal of Rare Earths. 2010;**28**:171-173

[70] Zhang N, Chen D, Niu F, Wang S, Qin L, Huang Y. Enhanced visible light photocatalytic activity of Gd-doped $BiFeO_3$ nanoparticles and mechanism insight. Scientific Reports. 2016;**6**:1-11

[71] Subha K, Kamath M, Kumar R. Delineating the photocatalytic properties

of doped mullite Bi2Fe4O9 by virtue of Gd3+ ions. Materials Letters. 2021;**297**:129960

[72] Iliev MN, Abrashev MV, Lee H-G, Popov VN, Sun YY, Thomsen C, et al. Raman spectroscopy of orthorhombic perovskite like YMnO3 and LaMnO. Physical Review B. 1998;**57**:2872

[73] Iliev MN, Lee H-G, Popov VN, Abrashev MV, Hamed A, Meng RL, et al. Raman- and infrared-active phonons in hexagonal YMnO3: Experiment and lattice-dynamical calculations. Physical Review B. 1997;**56**:2488

[74] Gibbs AS, Knight KS, Lightfoot P. High-temperature phase transitions of hexagonal YMnO3. Physical Review B. 2011;**83**:094111

[75] Zaghrioui M, Ta Phuoc V, Souza RA, Gervais M. Polarized reflectivity and lattice dynamics calculation of multiferroic YMnO3. Physical Review B. 2008;**78**:184305

[76] Glinsek S, Mahjoub MA, Rupin M, Schenk T, Godard N, Girod S, et al. Fully transparent friction-modulation haptic device based on piezoelectric thin film. Advanced Functional Materials. 2020;**30**:2003539

[77] Sharath SU, Vogel S, Molina-Luna L, Hildebrandt E, Wenger C, Kurian J, et al. Control of switching modes and conductance quantization in oxygen engineered HfOx based Memristive devices. Advanced Functional Materials. 2017;**27**:1-13

[78] Balamurugan C, Lee DW. Perovskite hexagonal YMnO3 nanopowder as p-type semiconductor gas sensor for H2S detection. Sensors and Actuators B: Chemical. 2015;**221**:857-866

[79] Van Aken BB, Palstra TTM, Filippetti A, Spaldin NA. The origin of ferroelectricity in magnetoelectric YMnO3. Nature Materials. 2004;**3**:164-170

[80] Aikawa Y, Katsufuji T, Arima T, Kato K. Effect of Mn trimerization on the magnetic and dielectric properties of hexagonal YMnO3. Physical Review B. 2005;**71**:184418

[81] Ito D, Fujimura N, Yoshimura T, Ito T. Influence of Schottky and Poole–Frenkel emission on the retention property of YMnO3-based metal/ferroelectric/insulator/semiconductor capacitors. Journal of Applied Physics. 2003;**94**:4036

[82] Kozlenko DP, Kichanov SE, Lee S, Park JG, Glazkov VP, Savenko BN. High-pressure effect on the crystal and magnetic structures of the frustrated antiferromagnet YMnO3. Journal of Experimental and Theoretical Physics Letters. 2005;**82**(4):193-197

[83] Muñoz A, Casáis MT, Alonso JA, Martínez-Lope MJ, Martínez JL, Fernández-Díaz MT. Complex magnetism and magnetic structures of the metastable HoMnO3 perovskite. Inorganic Chemistry. 2001;**40**:1020-1028

[84] Singh AK, Jain D, Ganesan V, Nigam AK, Patnaik S. Field-dependent competing magnetic ordering in multiferroic Ni3V2O8. Europhysics Letters. 2009;**86**:57001

[85] Abkenar GN, Viricelle J-P, Rieu M, Breuil P. Study of two vanadium based materials as working electrode for developing a selective mixed-potential ammonia sensor. Multidisciplinary Digital Publishing Institute Proceedings. 2018;**2**:770

[86] Yan A-L. Gas sensing properties of nickel vanadate and composite of nickel vanadate/Polypyrrole. Sensors Mater. 2018;**30**:1277-1282

[87] Liu F, Sun R, Guan Y, Cheng X, Zhang H, Guan Y, et al. Mixed-potential type NH3 sensor based on stabilized zirconia and Ni3V2O8 sensing electrode. Sensors and Actuators B: Chemical. 2015;**210**:795-802

[88] Inaudi D, Glišić B. Integration of Distributed Strain and Temperature Sensors in Composite Coiled Tubing. San Diego, CA, USA: SPIE Smart Structures and Materials Conference; 2006

[89] Inaudi D, Glišić B. Long-range pipeline monitoring by distributed fiber optic sensing. In: Proc. IPC2006, 6th Int. Pipeline Conf. Calgary, Al, Canada; 2006

[90] Atkins RA et al. Fiber-optic pressure sensors for internal combustion engines. Applied Optics. 1994;**33**(7):1315-1320

[91] Sadkowski R et al. Multiplexed interferometric fiber-optic sensors with digital signal processing. Applied Optics. 1995;**34**(25):5861-5866

[92] Bae T et al. Interferometric fiber-optic sensor embedded in spark plug for in-cylinder pressure measurement in engines. Applied Optics. 2003;**42**(6):1003-1007

Chapter 4

Toxic Gas Detectors Based on a $MnSb_2O_6$ Oxide Chemical Sensor

José Trinidad Guillen Bonilla, Héctor Guillen Bonilla, Maricela Jiménez Rodríguez, Alex Guillen Bonilla, Verónica María Rodríguez Betancourtt, Víctor Manuel Rangel Cobian, María Eugenia Sánchez Morales and Antonio Casillas Zamora

Abstract

We synthesized the semiconductor oxide $MnSb_2O_6$ through a wet chemical process assisted by low-power microwave radiation. A gas-sensitive sensor was elaborated from the $MnSb_2O_6$ powders obtained by calcination at 600°C. The sensor was electrically characterized in static CO and C_3H_8 atmospheres by measuring direct current signals at 100, 200, and 300°C. The toxic gases' concentrations were 1, 5, 50, 100, 200, 300, 400, and 500 ppm of C_3H_8; and 1, 5, 50, 100, 200, and 300 ppm of CO. From the $MnSb_2O_6$'s electrical resistance results, a sensor's operational point and a low-cost analog circuit were proposed, obtaining two new prototypes: one for detecting C_3H_8 and a second one for detecting CO. We selected the response at 200°C and 5 ppm for both cases. Notably, this concentration (5 ppm) is selectable with a calibration resistance, generating an alarm signal of $\approx 11.3\ V$ at a supply voltage of 120 V AC. The toxic gas detectors showed excellent functionality. The resistive sensor showed high sensitivity and good electrical response, while the analog circuit presented a rapid response. Due to the operating temperature employed (200°C), these devices could find practical applications, for example, exothermic generators and heaters.

Keywords: semiconductor oxide $MnSb_2O_6$, sensor was electrically characterized, good electrical response, resistive sensor, CO and C_3H_8 atmospheres, toxic gas detectors, analog circuit

1. Introduction

The detection of polluting gases is of great importance worldwide because an alarming increase in respiratory, ocular, skin, and lung diseases has been detected [1, 2]. There is no adequate control on the emission of toxic gases into the atmosphere by many large industries and transportation systems [2]. Therefore, a huge effort in research and synthesis of new materials capable of detecting different levels of toxic gases has been undertaken. The current aim is the development of new inexpensive, reliable, sensitive, efficient, and thermally stable detectors in toxic atmospheres (like in CO, CO_2, C_3H_8, NO_2, etc.) [1–4]. The most widely studied semiconductors for their application as gas detectors are the p and n-type binary oxides (like SnO_2, ZnO, NiO, CuO, In_2O_3, WO_3, Fe_2O_3, etc.) [4–6], as well as the ternary perovskite-type oxides (like $YCoO_3$ and $LaFeO_3$) [7, 8], and the oxides with spinel-type structure (like $CoFe_2O_4$ and $ZnFe_2O_4$) [9, 10]. However, recent reports informed that other transition-metal ternary semiconductors with a trirutile-type crystalline structure (AB_2O_6, where A and B possess divalent and pentavalent bonds, respectively) are alternative materials for toxic-gas sensors [11, 12]. Among these materials are the $CoSb_2O_6$ [11], the $NiSb_2O_6$ [12], the $MgSb_2O_6$ [13], and the $ZnSb_2O_6$ [14]. These oxides have been studied in CO_2, CO, C_3H_8, LPG, H_2S, and NO_2 atmospheres [11–17].

The $MnSb_2O_6$ oxide (manganese antimonite, where in this case the Mn ion substitutes the A) can show a hexagonal crystalline structure with the spatial group P_{321} [18] or a trigonal chiral-type crystalline structure [19]. The $MnSb_2O_6$ possesses good catalytic activity in toxic atmospheres (such as CO) at temperatures above 100°C [20, 21]. In Ref. [20], it was reported that the $MnSb_2O_6$ showed a maximum sensitivity magnitude (S) of ~8.98 at 300 ppm of CO and 300°C. At 500 ppm of C_3H_8 (propane), the response was ~0.439 at the same temperature. In Ref. [21], the $MnSb_2O_6$ was synthesized by the colloidal method and it showed increased sensitivity as a function of the increase in the operating temperature and the gases' concentrations. The response in 500 ppm of propane had a maximum of ~165.66 at 300°C. In 300 ppm of CO, the maximum response was ~14.929, also at 300°C. The good $MnSb_2O_6$'s electrical response was attributed to the nanometric particle's microstructure (porosity, morphology, and size) obtained during the synthesis process.

We studied the $MnSb_2O_6$ regarding its application as a gas detector and found little information on the application of the oxide in devices for detecting C_3H_8 and CO, which form atmospheres with a high risk of explosion (propane) or intoxication (CO). In this work, we designed two novel gas detectors: one of them detects C_3H_8 concentrations and the second one detects CO concentrations. Both are analog detectors and were designed based on the oxide's electrical response. They are capable to detect concentrations of 5 ppm, which can be modified through a calibrating resistance. They generate an alarm signal at ~ 11.3 V, use a supply voltage of 120 V in alternate current (AC), and an operating voltage of $V_{cc} = 12$ V. They possess rapid response, and are easy to build, install, and repair. Both are ideal for industrial applications where combustion is involved.

2. Materials and methods

2.1 Synthesis of powders $MnSb_2O_6$

$MnSb_2O_6$ powders were synthesized by the wet chemistry method assisted by microwave radiation reported in the literature [21]. In particular, for this work, a synthesis of the oxide ($MnSb_2O_6$) at room temperature, in presence of ethylenediamine was developed. For the synthesis of the $MnSb_2O_6$ 5 mmol of Mn $(NO_3)_2$•$4H_2O$ (Sigma-Aldrich), 10 mmol of $SbCl_3$ (Sigma-Aldrich), 8 mmol of ethylenediamine (Sigma-Aldrich), and ethyl alcohol (Golden Bell) were used. These were dissolved separately in 5 ml of ethyl alcohol, except for ethylenediamine, which was added to 10 ml of the same solvent. The three solutions obtained were transparent

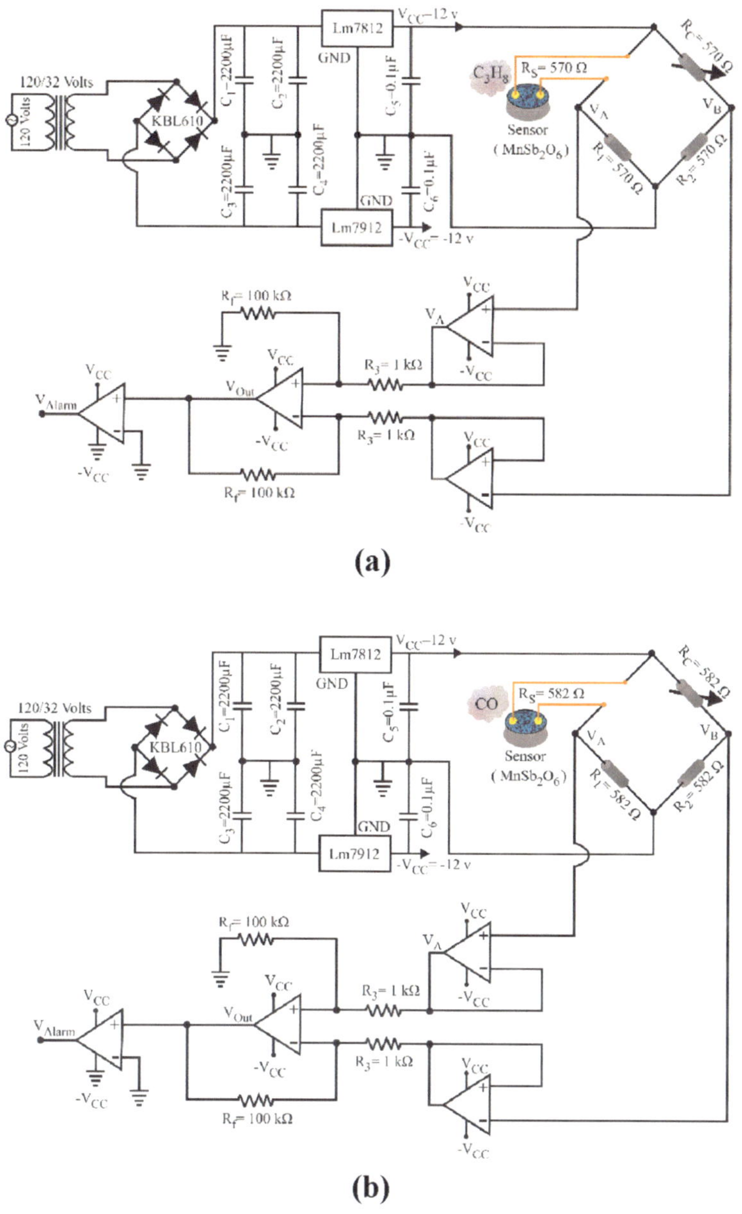

Figure 1.
Electronic diagrams: a) propane detector; b) carbon monoxide detector.

and stirred for a period of 30 min. Subsequently, the solutions with ethylenediamine were added dropwise to the solution with manganese nitrate without stopping stirring. Then, antimony chloride was also added dropwise to the mixture formed, obtaining a white solution with very fine particles dispersed throughout the solution (colloidal dispersion). The resulting solution was left stirring under stirring for 24 h at room temperature (at 370 rpm). The evaporation of the solvent from this solution was carried out by applying microwave radiation, using a General Electric model JES769WK domestic device operating at a power of 130 W. The microwave

Materials for the propane detector					
Gas	**Reactants**	**Wheatstone bridge**	**Instrumental circuit**	**Comparator**	**Supply source**
Propane (C_3H_8)	1.255 g of $Mn\ (NO_3)_2$•$4H_2O$ (Sigma-Aldrich) 2.28 g of $SbCl_3$ (Sigma-Aldrich) 0.5 mL de ethylenediamine 20 mL of ethyl alcohol (Golden Bell)	$R_c = 570\ \Omega$ $R_1 = 570\ \Omega$ $R_2 = 570\ \Omega$	3 Operating amplifiers 2 resistances $R_3 = 1\ K\Omega$ 2 resistances $R_f = 100\ K\Omega$	1 Operating amplifier	1 transformer 120/32 1 diode bridge KBL610 1 regulator LM7812 1 regulator LM7912 4 capacitors $C_1 = C_2 = C_3 = C_4 = 2{,}200\ \mu F$ 2 capacitors $C_5 = C_6 = 0.01\ \mu F$
Materials for the carbon monoxide detector					
Carbon monoxide (CO)	1.255 g of $Mn\ (NO_3)_2$•$4H_2O$ (Sigma-Aldrich) 2.28 g of $SbCl_3$ (Sigma-Aldrich) 0.5 mL of ethylenediamine 20 mL of ethyl alcohol (Golden Bell)	$R_c = 582\ \Omega$ $R_1 = 582\ \Omega$ $R_2 = 582\ \Omega$	3 Operating amplifiers 2 resistances $R_3 = 1\ K\Omega$ 2 resistances $R_f = 100\ K\Omega$	1 Operating amplifier	1 transformer 120/32 1 diode bridge KBL610 1 regulator LM7812 1 regulator LM7912 4 capacitors $C_1 = C_2 = C_3 = C_4 = 2{,}200\ \mu FF$ 2 capacitors $C_5 = C_6 = 0.01\ \mu F$

Table 1.
Materials employed for the construction of the C_3H_8 *and CO detectors.*

applications made to the solution were made in steps of 60 s until reaching a time of 160 min. The precursor material obtained from evaporation (a white paste) was heated at 200°C (drying process) for 8 h and later it was calcinated at 600°C with increments of 100°C/h for 5 h in static air. The thermal treatment applied to the composite was carried out with a muffle (Novatech) with programmable control.

2.2 Characterization equipment

To know the purity and crystallinity of the $MnSb_2O_6$,X-ray diffraction was used in the powder. For this study, a Panalytical Empyrean equipment with CuKα radiation and a wavelength λ of 1.540598 Å by a continuous 2θ scanning from 10 to 70°, with 0.026°-steps at a rate of 1 second per step. The microstructure of the powders of the compound at the micrometric scale was analyzed with field emission scanning electron microscopy (FE-SEM) that uses a system Tescan MIRA 3 LMU with an acceleration voltage of 10 kV in a high vacuum. To determine the morphology and particle size at the nanometric scale, a transmission electron microscope was used (TEM), model JEM-2100 with an acceleration voltage of 200 kV, in image mode. It is important to mention that to study the individual nanoparticles by TEM, the powders of the compound (0.01 g) were placed inside a vial that was previously provided with 1 ml of ethyl alcohol. The vial with the powders was dispersed with an ultrasonic generator for 5 min. Later a drop with the material was extracted and deposited on a 300-mesh copper grid that had a Formvar/carbon membrane.

2.3 Materials

In **Figure 1a**, an electronic diagram is depicted of the propane gas detector. In **Figure 1b**, the carbon monoxide detector is shown. As can be noted, the devices consist of a Wheatstone bridge for adapting the electric signal of the gas sensor, a circuit based on three operating amplifiers for performing a comparison between the Wheatstone bridge exit signals and an instrumental circuit for amplifying the signal generated by the Wheatstone bridge when the sensor detects the presence of the toxic gas. The complete list of materials is shown in **Table 1**.

As can be seen in **Table 1**, both detectors employ nearly the same materials. The only difference is the Wheatstone bridge resistances because the gas sensor produces different resistive responses for each gas.

2.4 Process

In **Figure 2**, we show schematically the construction of the gas detectors. It consisted of three stages: 1) fabrication and characterization of the gas sensor, 2) analog electronic circuit for the adaptation, and 3) voltage source for the electronic circuit and the sensor.

In stage 1, the resistive sensor consisted of pellets made with the oxide. It was exposed to different gas concentrations and three operating temperatures: 100, 200, and 300°C. The resistivity was measured with a digital multimeter. In stage 2, we performed a signal adaptation of the gas sensor using a Wheatstone bridge, an instrumental circuit, and a comparator circuit. The exit signal was an alarm signal produced when the system detected the presence of a gas. In stage 3, the supply source had an exit voltage of $V_{cc} = \pm 12$ Volts for the sensor, the Wheatstone bridge, the instrumental circuit, and the comparator circuit.

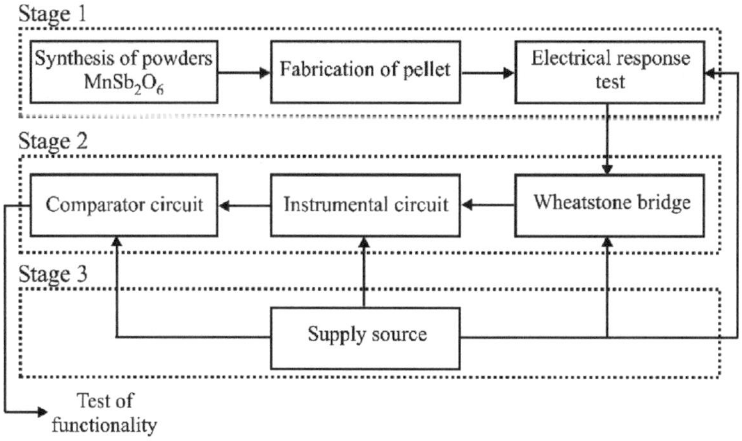

Figure 2.
Schematic diagram for the fabrication of the toxic gas detectors.

2.5 Stage 1: Sensor construction and electrical characterization

We elaborated pellets from $MnSb_2O_6$ powders calcinated at 600°C. For that, we weighed 0.3 g of $MnSb_2O_6$ powder and deposited them in a circuit breaker box. The breaker box was placed in a hydraulic press (Simplex Ital Equip-25-ton-brand equipment) and applied 10 tons of pressure for 1 min. The dimensions of the pellets were 12 mm in diameter and 0.5 mm in thickness. We placed two colloidal-silver paint (Alfa Aesar ˃ 99%) ohmic contacts on the pellets' surface to maintain a good connection between the electrodes and the pellets' surface. The pellets were placed inside a high vacuum chamber with a capacity of 10^{-3} Torr. The test gases were injected individually and then removed from the chamber with a vacuum pump. The gas concentration and the partial pressure were constantly monitored and controlled with a Leybold detector (model TM20). The changes in the electrical resistance were recorded with a digital multimeter (Keithley 2001) as a function of operating temperature (100–300°C) and gas concentration (CO: 1–300 ppm, C_3H_8: 1–500 ppm). Pellets' resistivity was calculated with equation [20]:

$$\rho = \frac{RA}{t} \tag{1}$$

where R is the pellets' electrical resistance in the tested gas, A is the pellets' area (12 mm in diameter), and t is the thickness (0.5 mm).

2.6 Stage 2: Gas sensor's signal adaptation

Electronic circuits, comprising a Wheatstone bridge, an instrumental circuit, and a comparator circuit, were designed (**Figure 1**) to generate an alarm signal when the resistive sensor detected a gas. These circuits were supplied by a $V_{cc} = \pm 12V$ voltage source, which is described in Section 2.6.1. Whereas, in Section 2.6.2., we describe the functioning of the gas detector.

2.6.1 Voltage source

We were interested in developing toxic gas detectors based on the antimoniate manganese oxide and analog electronic circuits from an economic perspective (they should not be very expensive). Thus, we developed a voltage source for feeding our prototypes. Its electronic diagram is shown in **Figure 3**. The functioning of each stage is as follows:

Voltage reduction is done with the 120 to 32 V reducer transformer. When the transformer is connected to the 120 V alternate current signal, the primary coil generates a magnetic field that reaches the secondary coil. According to the number of wire rotations in the secondary coil, the transformer generates an output of 32 V.

Alternate current signal rectification is done by a KBL610 diode bridge, which consists of four rectifier diodes. This circuit converts the alternate current signal into a direct current signal. However, the diode bridge's signal contains very loud noise components, thus requiring a third filtering stage.

Filtering of the signal is done through capacitors C_1, C_2, C_3, and C_4, which should possess a sufficiently large capacitance value to eliminate to a certain degree the voltage of the curling produced by the alternate current rectification.

Voltage regulation by the regulators LM7812 and LM7912. LM7812 is a 12 V positive voltage regulator and LM7912 is a $-$ 12 V negative voltage regulator. Both regulators require 0.1 μF capacitors (C_5 and C6, respectively).

The voltage source supplies the device using a linear behavior. Its construction is easy and very cheap.

2.6.2 Operating principle

The operation consists of two stages: Calibration and detection. During the calibration, the device does not detect the presence of gas and the alarm signal is set to zero, $V_{Alarm} = 0$. To achieve this, we followed these steps: a) the resistive sensor was placed in an atmosphere free of the test gas, and its terminals were connected to a Wheatstone bridge's arm; b) the variable resistance R_c changed its value until the Wheatstone bridge's exit voltage was equal to zero:

$$V_A - V_B = \frac{R_1}{R_1 + R_s} V_{cc} - \frac{R_2}{R_2 + R_c} V_{cc} = 0 \tag{2}$$

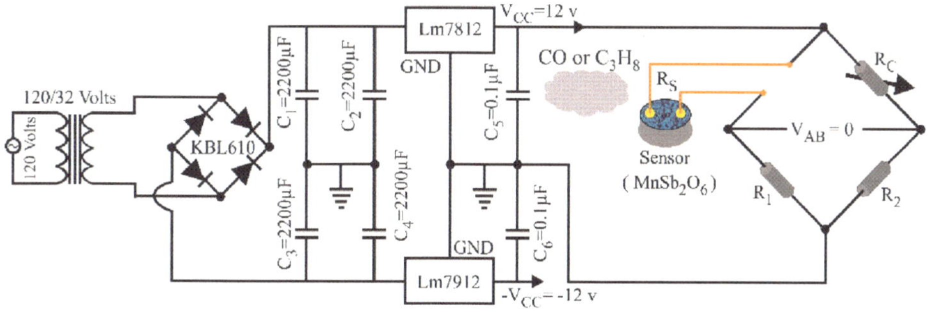

Figure 3.
Resistive sensor's connection to the Wheatstone bridge and diagram of the voltage source.

where R_1 and R_2 are the precision resistances, R_s is the sensor's resistance at initial conditions, R_c is the variable resistance, and V_{cc} is the supply voltage; c) V_A and V_B are simultaneously the Wheatstone bridge exit voltages, and the entry and exit voltages for the follower circuits (see **Figure 1**); d) the adder-subtracter yields the difference between V_A and V_B, and amplifies the entry signal based on the ratio $\frac{R_f}{R_3}$:

$$V_{OUT} = \frac{R_f}{R_3}(V_A - V_B) = 0. \quad (3)$$

If $V_A - V_B = 0$ (see Eq. 2), the voltage V_{OUT} is also equal to zero; and e) the exit signal of the comparator circuit is the device's alarm signal and is determined by

$$V_{Alarm} = V_{Sat} = A_{ol} V_{\text{OUT}} = 100000(0) = 0 \quad (4)$$

where $A_{ol} \approx 100000$ is the gain of the operating amplifier's open loop. In this case, the detector does not generate the alarm signal ($V_{Alarm} \approx 0$) because the sensor does not detect the gas. In the detection phase, the device detects the gas. The alarm signal is equal to the operating amplifier's saturation voltage. In this stage, we followed these steps: a) the resistive sensor was placed in an atmosphere for monitoring the possible presence of the test gas; b) the sensor's terminals were connected to a Wheatstone bridge's arm (see **Figure 1**); c) when the device detected the gas, the Wheatstone bridge was not balanced, satisfying following condition:

$$V_A > V_B = \frac{R_1}{R_1 + R_s - \Delta R_s} V_{cc} > \frac{R_2}{R_2 + R_c} V_{cc}, \quad (5)$$

where ΔR_s is the sensor's resistance variation due to the presence of the gas, the adder–subtractor circuit's exit voltage is then greater than zero:

$$V_{OUT} = \frac{R_f}{R_3}(V_A - V_B) > 0; \quad (6)$$

and d) if $V_{OUT} > 0$, the comparator circuit's amplifier is saturated and the device's exit voltage is the alarm signal $V_{\text{Alarm}} = V_{\text{Sat}}$, that is, the device has detected the presence of toxic gas in the atmosphere.

3. Experimental results

3.1 XRD analysis

Once the experimental process to obtain the powders of the $MnSb_2O_6$ through a wet chemistry method, in **Figure 4a** diffractogram of $MnSb_2O_6$ powders calcinated at 600°C is shown. In this X-ray diffraction pattern, it is identified that the largest peaks belong to the crystalline phase of $MnSb_2O_6$. These high-intensity reflections were compared taking as reference the file PDF # 84–1237 of the database. According to this file, the oxide $MnSb_2O_6$ presents a crystalline hexagonal structure with spatial group P321 [18, 22] and cell parameters a = 8.8054 Å and c = 4.7229 Å [19, 20]. According to these values and the major diffraction peaks shown in **Figure 4**, this phase corresponds to materials known in the literature as type antimonates ASb_2O_6 [23, 24]

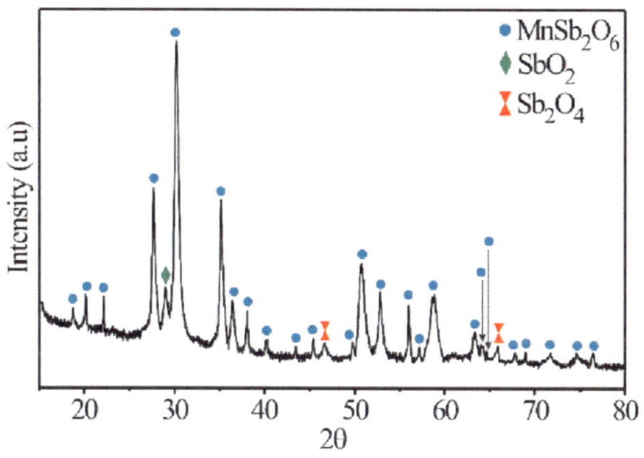

Figure 4.
Diffractogram of the $MnSb_2O_6$ oxide (at 600°C) prepared by a microwave radiation-assisted wet chemistry process.

where A is a divalent ion as Co, Zn, Mi, and Mn (Mn, in our case) [20, 24]. In addition, it is observed that the width and height of the peaks presented by the diffractogram of $MnSb_2O_6$ indicate good purity [20], and the relatively low noise shows a high crystallinity [25, 26]. However, even after having achieved a good crystallinity of the oxide at 600°C, small portions of the inorganic material SbO_2 (PDF#11–0694) and Sb_2O_4 (PDF#78–2067) were identified, which agree with what was reported in the reference [20, 21], in which the same ($MnSb_2O_6$) was prepared.

In the literature, the preparation of the oxide $MnSb_2O_6$ has been reported using the solid state reaction method at a temperature of 900°C and 1100°C [27]. While in Ref. [20], they mention that they synthesized $MnSb_2O_6$ at a temperature of 800°C applying a colloidal method. These temperatures reported by these authors are high compared to those obtained in this work, which was able to obtain the crystalline phase of $MnSb_2O_6$ at 600°C by using an alternative method of synthesis.

3.2 SEM analysis

In **Figure 5**, scanning electron microscopy (SEM) images of the surface of the calcined $MnSb_2O_6$ oxide at 600°C are shown. To analyze the microstructure of the oxide in detail, it was necessary to use the following three magnifications: (a) 3.05kx, (b) 5.72kx, and (c) 9.72kx, respectively. Regarding the low magnification of 3.05kx (**Figure 5a**), we observe a large number of individual microrods, constituted by the agglomeration of very fine and irregular particles. These microrods appear to grow dispersedly oriented in different directions until they form micro bases of different sizes. The length of the microrods was calculated in the range of 1 to 6 μm, with an average of ~2.8527 μm and a standard deviation of ~0.8770 μm (**Figure 6a**).

In agreement with the literature, the growth of these microcolumns is associated with the increase in temperature and the effects that ethylenediamine produces on the morphology of the compound [20, 28, 29]. In addition, in this same image, plates composed of irregular and very fine particles can also be seen. At a magnification greater than 5.72kx (**Figure 5b**), we corroborate that the microrods obtained at a temperature of 600°C are composed of the agglomeration of irregular and very small particles. The average diameter of the microrods analyzed was estimated to range

Figure 5.
SEM photomicrographs of the $MnSb_2O_6$ powders analyzed at magnifications of (a) 3.05kx, (b) 5.72kx, and (c) 9.72kx.

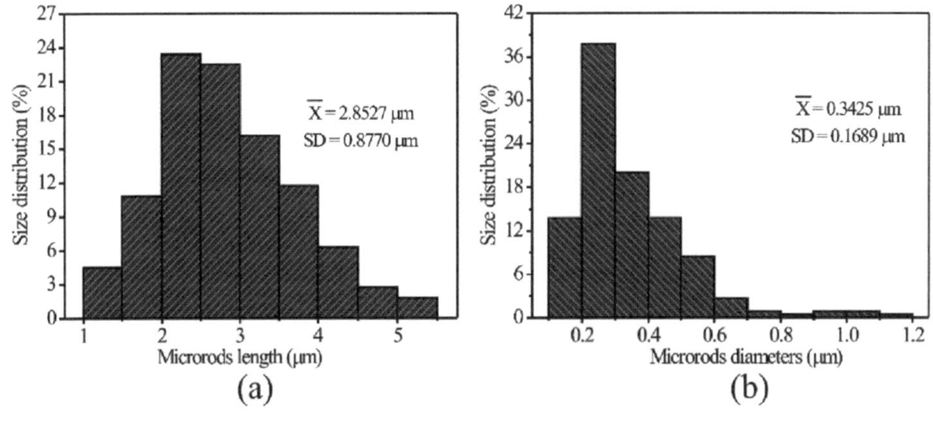

Figure 6.
Histograms of the distribution of microrod-like morphologies of $MnSb_2O_6$ calcined at 600°C, (a) microrod length distribution and (b) microrod diameter distribution.

from 0.1 to 1.2 μm, with a mean of ~0.3425 μm and a standard deviation of ~0.1689 μm (**Figure 6b**). The granular surface shown by the microcolumns in **Figure 5b** is due to the removal of organic material that was present during the oxide synthesis process.

Observing another zone of the surface of the material, **Figure 5c** shows an image analyzed at a magnification of 9.72kx. According to the study carried out in this area of the material, particles with different morphologies to those described in **Figure 5a** and **b** were found. In this case, a particle that grows with an octahedral geometry (with a size of 1.74 μm) was identified. In the same image, the growth of microrods and very fine particles can be observed that, when agglomerated by the effect of the calcination temperature, form micro bases where microcolumns and other particles without apparent shape grow. The morphologies described in **Figure 5c** are formed due to the fact that these microstructures are nucleated, taking as raw material the finest particles that are found surrounding the microrods [29, 30].

In the literature, results like those shown in **Figure 5a-c** have been reported. For example, Michel et al. synthesize the $CoSb_2O_6$ oxide using the colloidal route and using low concentrations of ethylenediamine (0.5 mL) [31]. According to the results reported by this author, he obtains the formation of microcolumns and other irregular particles due to the use of ethylenediamine during the synthesis of the compound [31]. In Ref. [20], they prepare the oxide $MgSb_2O_6$ applying the colloidal method in the presence of ethylenediamine (4 mL). In this report, they mention that with the use of ethylenediamine in the preparation of trirutile-type oxides, microrods composed of irregular nanoparticles are obtained and that they agglomerate due to the effect of temperature until elongated morphologies (microrods) are obtained [13]. The authors cited above apply these microstructures for their study as potential gas sensors [20, 21]. In our case, it was possible to synthesize the $MnSb_2O_6$ by a simple, economical, and easy-to-control wet chemical process in the presence of ethylenediamine, obtaining microrods, and other irregular particles for their study as potential sensors of C_3H_8 and CO atmospheres. **Figure 6** shows the measured particle size.

3.3 TEM analysis

In order to have a clearer idea of the individual morphology and to make a more precise estimation of the particle size of the material, four TEM images (one of them in high resolution; HRTEM) acquired from the surface of the $MnSb_2O_6$ oxide are shown in **Figure 7**. calcined at 600°C. The dark areas observed in these photomicrographs are due to the poor transmission of electrons to pass through the thickness of the particles. Looking at **Figure 7a**, we can see the formation of a large agglomeration of dispersed particles on the surface of the material. The scattering of the particles that are observed in this image is due to the fact that, in the process of preparation for its study by TEM, the $MnSb_2O_6$ powders were dispersed with the purpose of analyzing the individual particles and giving the appearance shown in **Figure 7a**. The average size of these particles was calculated as an average of 150 nm, with a range of 100 to 220 nm. For **Figure 7b**, elongated morphologies were found and identified as microrods. In this case, the studies carried out by scanning electron microscopy (SEM) are confirmed, where the growth of microrods was recorded (see, **Figure 5**). The length of the microrod in **Figure 7b** was estimated as 438 nm in length and 140 nm in diameter. In **Figure 7c**, an individual particle is presented, with this image, we also corroborate that the $MnSb_2O_6$ powders are made up of irregular nanoparticles (see also **Figure 7a**).

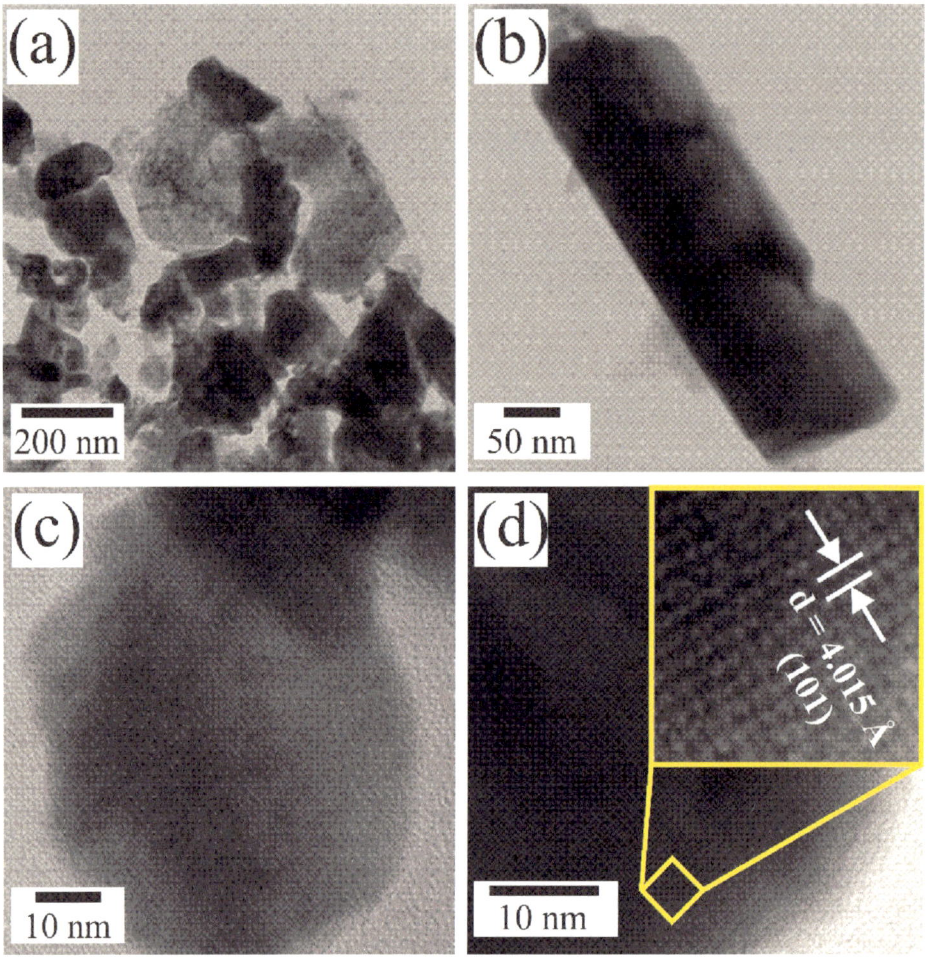

Figure 7.
TEM images of $MnSb_2O_6$ powders calcined at 600°C showing (a) particle dispersion, (b) microrod growth, (c) individual morphology of a nanoparticle, and (c) a nanoparticle where observes the crystalline planes of the compound observed in high resolution (HRTEM).

The size of the analyzed nanoparticle was estimated to be approximately 55 nm in size. Referring to **Figure 7d**, a high-resolution TEM (HRTEM) image of the surface of an individual nanoparticle is shown. Of this particle, an enlargement (zoom) of the surface was made in order to identify the crystalline planes formed by the effect of the calcination temperature (600°C). According to this analysis, the presence of the crystalline planes of the compound is observed, indicating its crystalline nature. The distance between the (d) planes was calculated to be 4.01 Å, which corresponds to the (101) plane of the hexagonal structure. The diffraction angle for the found plane is 2θ = 22.12°. These results can be verified in the diffractogram of **Figure 4**.

3.4 Electrical response

The electrical resistivity variations of $MnSb_2O_6$ pellets in CO and C_3H_8 atmospheres, at the operating temperatures of 100, 200, and 300°C, and concentrations of 1, 5, 50, 100, 200, and 300 ppm of CO and 1, 5, 50, 100, 200, 300, 400, and 500 ppm of C_3H_8, are depicted in **Figure 8** as resistivity (ρ) vs. test gas concentration and operating temperature. As can be observed in **Figure 8a** and **b**, at 100°C, no changes in electrical resistivity were detected, regardless of the increase in gas concentration. That is because, at this temperature, the thermal energy is not enough to drive a reaction of CO or C_3H_8 with the pellets' surface [13], causing poor mobility of the charge carriers (in this case, electrons) on the surface [20]. Additionally, that temperature causes the oxygen adsorption and desorption processes not to take place [26]. According to the literature, when a semiconductor is employed as a gas sensor in atmospheres similar to those studied in this work, at temperatures below 150°C, the available oxygen species are of type O_2^- [8, 11, 13, 20], which are low reactive at such temperatures. Therefore, the oxygen desorption process does not take place, regardless of the increase in CO and C_3H_8 concentrations [13, 20, 21]. On the other hand, by raising the operating temperature to 200 and 300°C, the electrical resistivity diminishes with the increasing concentration and operating temperature. The decrease in both atmospheres was more evident at 300°C (see **Figure 8b** and **d**). That is attributed to the fact that, as the temperature increases from 100 to 300°C, the dynamic activity of the charge carriers rises [11, 12, 14], contributing to the increase of the material's conductivity and the adsorption and desorption processes [4, 14, 21, 32] at 200 and 300°C. The excellent pellets' response at those temperatures is due to the strong

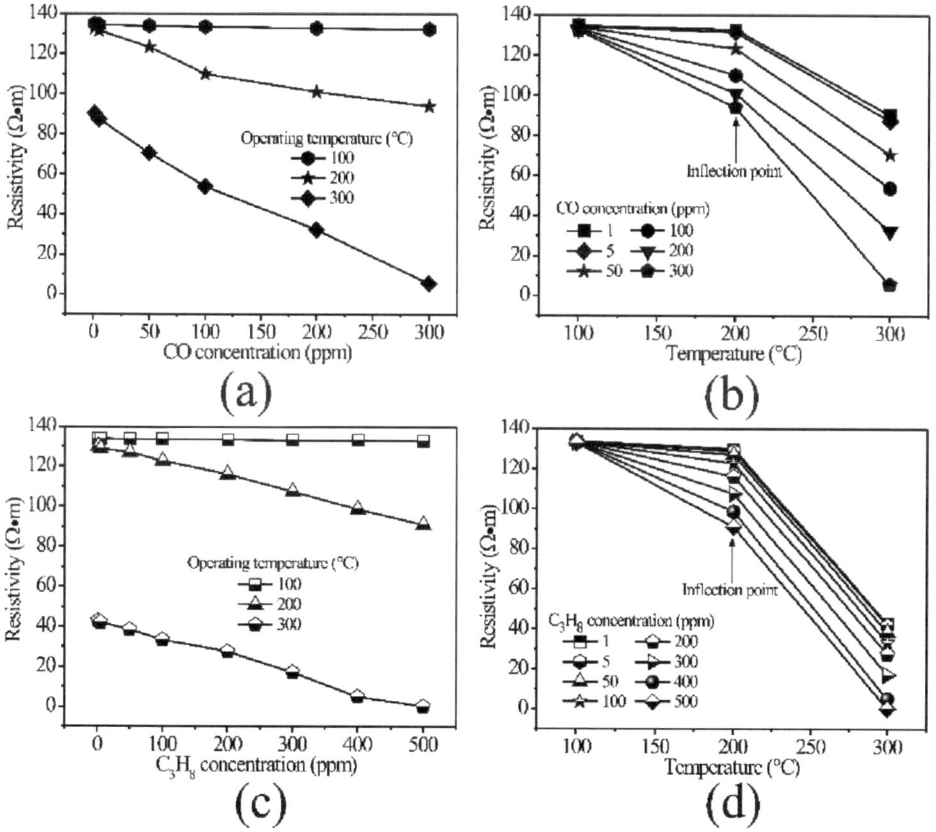

Figure 8.
Electrical response of $MnSb_2O_6$ pellets as a function of (a, b) CO and C_3H_8 concentrations, and (c, d) operating temperature.

chemical reaction of the oxygen species present on the pellets' surface and to the increased concentration of the gases [12, 26]. We observed that when the temperature increased from 100 to 200°C, the electrical resistivity variations were more abrupt, as can be observed in **Figure 8c** and **d**. An inflection point, due to the high diffusion of the gases that reacted with the oxygen can be seen [20, 21, 32]. Also, we corroborated that the best pellets' response at 300°C was due, in great measure, to the increase of the thermal energy, which provoked better oxygen species' ($O^-$and O^{2-} -ionic forms) absorption on the surface [8, 13, 21, 26], thus causing higher velocity of the charge carriers and leading to an increase in the material's conductivity (or decreased resistivity) [12, 31]. In agreement with the literature, the different oxygen species (such as O_2^-,O^-,or O^{2-}) that react at temperatures above 150°C [8, 26, 33] are the most probable responsible for causing a rise of the electrical response as a function of the increase in gas concentration and operating temperature (see **Figure 8a-d**). The values of the electrical resistivity at 300°C were 90.40, 87.46, 70.51, 53.78, 32.31and 5.67 Ωm for CO and 42.71, 42.03, 38.42, 33.44, 27.57, 17.24, 5.16 and 0.25 Ωm for C_3H_8. This trend, depicted in **Figure 8a-d**, is normally shown in semiconductor oxides employed as toxic-gas sensors.

3.5 Gas-detection devices

The Wheatstone bridge previously discussed was calculated when the sensor's resistance had a value of $R_s = 570\ \Omega$ (in C_3H_8) and $R_s = 582\ \Omega$ (in CO). These resistance values were calculated using the Eq. 1: $\rho = \frac{RA}{t} \rightarrow R = \frac{\rho t}{A}$ and the experimental measurements were shown in **Figure 8a** and **c**. Now, according to **Figures 1, 3**, and **8**, it is possible to establish three operating parameters for our devices: an operating temperature of 200°C, an operating gas concentration of 5 ppm, and the initial sensor's resistance R_s, mentioned above. On the other hand, another three operating parameters can be established for the electronic circuits: a supply voltage of 120 V AC, an operating voltage $V_{cc} = \pm 12V$,and an exit signal (or alarm signal) $V_{Alarm} \approx 11.3$ V. **Figure 9** shows our devices: **Figure 9a** depicts the C_3H_8 detector device, and **Figure 9b** shows the CO detector device.

Already manufactured the gas sensor and the electronic device according to the description of the previous sections, to apply the gas detection device (C_3H_8, CO), the

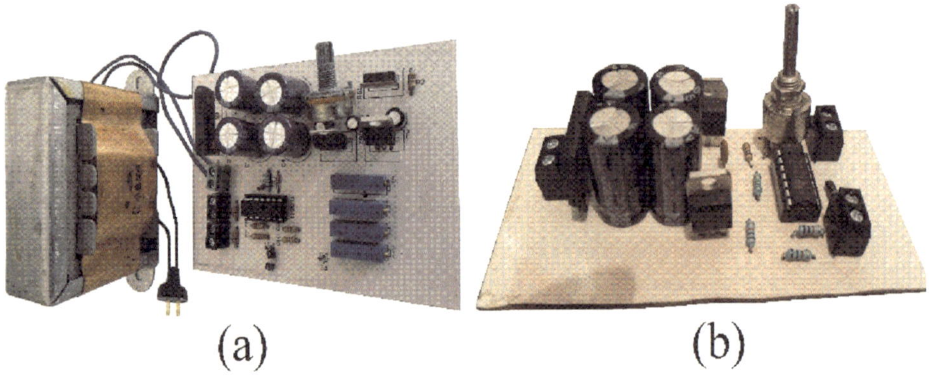

Figure 9.
Low-cost toxic gas detectors: a) for C_3H_8, b) for CO.

sensor must be installed in the atmosphere to be monitored whose temperature must be 200°C, while its terminals must be connected to a pair of terminals of the electronic card developed for the detection of toxic gas (C_3H_8 or CO). The electronic card is manufactured using a copper PCB with the size of 10 cm by 10 cm, its design was made with the program Proteus®, it operates at room temperature and its supply voltage is through a plug (see **Figure 5b**), which is connected to a 110 V alternating current voltage source. Finally, using a multimeter, the alarm signal voltage V_{Alarm} produced by the electronic card is measured: If the alarm voltage is $\approx 11.3V$, the device is detecting the presence of gas in the monitored atmosphere. But if $V_{Alarm} \approx 0V$ then the device is not detecting the presence of gas.

Figure 10a and **b** depict the operating principle of the C_3H_8 and CO detectors. It is as follows: If the concentration of the test gas is equal or greater than 5 ppm, the sensor's resistance diminishes, thus unbalancing the Wheatstone bridge, and causing the devices' exit voltage to be equal to that of the alarm signal $V_{Alarm} \approx 11.3V$ (alarm state: "On"). However, if the gas concentration is below 5 ppm, Eqs. (5) and (6) are not satisfied. Consequently, the alarm signal is not active (alarm state: "Off"), that is, the devices have an exit voltage of approximately zero V. It is worth mentioning that, for our devices, the operating threshold value is selectable with the variable resistance R_c. For the C_3H_8 detector, if $R_c > 570\ \Omega$ (see **Figure 1a** and **9a**), the device will detect concentrations lower than 5 ppm. On the other hand, for the CO detector, if $R_c > 582\ \Omega$ (see **Figure 1b** and **9b**), the device will detect concentrations lower than 5 ppm. If $R_c < 570\ \Omega$ or $R_c < 582\ \Omega$, respectively, the devices will be able to detect concentrations higher than 5 ppm.

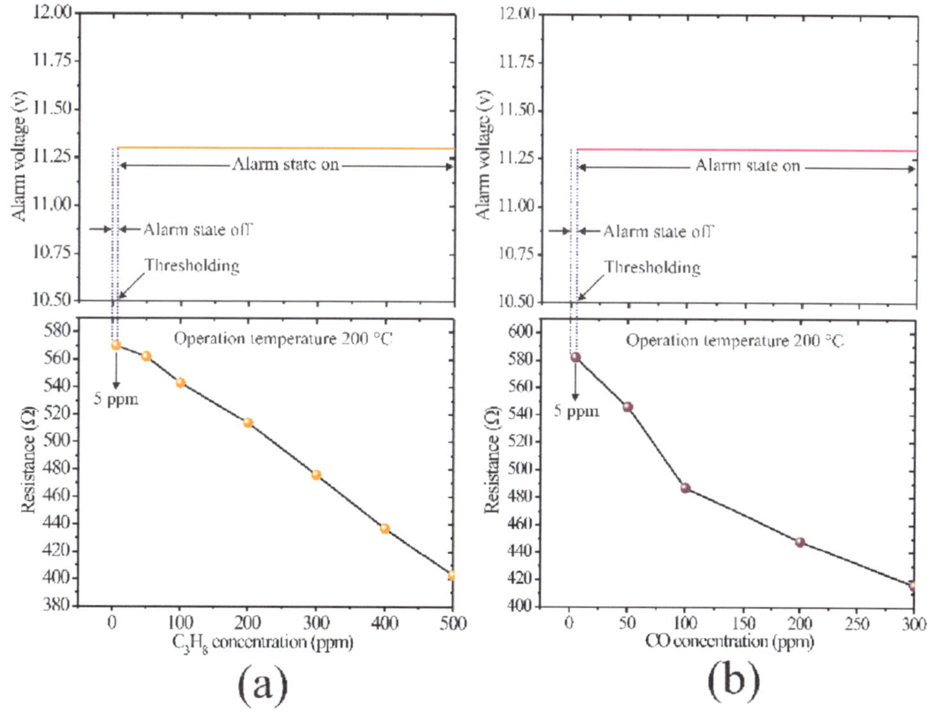

Figure 10.
Operating principle of our gas detectors: (a) propane detector and (b) carbon monoxide detector.

4. Discussion

According to our results, our C_3H_8 and CO-detector devices possess good features, which include low cost, high sensitivity, rapid response, good behavior, ability to select the operating concentration through a variable resistance, an operating temperature of 200°C, dimensions of 10 cm x 10 cm, a supply voltage of 120 V AC, an exit voltage of $V_{Alarm} \approx 11.3V$, ease of construction, ease of repair, and ease of use.

We previously proposed a CO_2 gas detector based on an analog circuit and on the dynamic response to the impedance of the oxide $CoSb_2O_6$ [22]. Such a device detected a gas concentration of 100 ppm with an operating temperature of 250°C. On the other hand, we also studied theoretically the dynamic electrical response of the oxide $ZnAl_2O_4$ and proposed a propane-gas detector [23]. That device detected gas concentrations of 1000 ppm with an operating temperature of 250°C. For both detectors, the design of the analog electronic circuits possessed high complexity since the analysis of the signal was conducted on complex planes and based on the sensors' dynamic response. In this work, the $MnSb_2O_6$ oxide was applied also in the detection of C_3H_8 and CO. However, its characterization and signal adaptation were done using DC currents, thus simplifying its analysis and implementation. Therefore, in comparison, our new proposal facilitates the construction of the device, lowering the operating temperature (from 250 to 200° C), the test-gas concentration threshold (from 100 ppm to 5 ppm for C_3H_8 and from 1000 ppm to 5 ppm for CO), and the device dimensions (from 15 cm x 15 cm to 10 cm x 10 cm). The overall components were also optimized.

Devices for the detection of C_3H_8 and CO find practical applications as explosion and intoxication-prevention measures, respectively. Our future work will be aimed at designing and developing gas detectors based on digital technology and quasi-distributed systems for the detection.

5. Conclusions

The reproducibility of the $MnSb_2O_6$ oxide was excellent. It was electrically characterized through static direct-current (DC) tests, obtaining its resistance-gas concentration behavior. Based on these results, the electronic prototypes for two toxic-gas-detection devices were designed: one of them for C_3H_8-detection, and the other one for CO-detection. Both prototypes have an operating temperature of 200°C and an operating concentration of 5 ppm. They can produce an alarm signal of approximately 11.3 V. Their supply and operating voltages are 120 V AC and $\pm$ 12 V, respectively. They possess fast response, ease of construction, ease of operation, very low cost, and ease of repair. The detectors can find rapid application in processes involving combustion like, for example, boilers, smelting furnaces, and exothermic generators.

Acknowledgements

The authors thank Mexico's National Council of Science and Technology (CONACyT) and the University of Guadalajara for their support. We also thank María de la Luz Olvera Amador and Jaime Santoyo Salazar for their technical assistance. This research was carried out following the research-line "Nanostructured Semiconductor Oxides" of the academic group UDG-CA-895 "Nanostructured Semiconductors" of CUCEI, University of Guadalajara.

Conflict of interest

The authors declare that there are no conflicts of interest regarding the publication of this article.

Author details

José Trinidad Guillen Bonilla[1]*, Héctor Guillen Bonilla[2], Maricela Jiménez Rodríguez[3], Alex Guillen Bonilla[4], Verónica María Rodríguez Betancourtt[5], Víctor Manuel Rangel Cobian[2], María Eugenia Sánchez Morales[6] and Antonio Casillas Zamora[1]

1 Departamento de Electro-fotónica, Centro Universitario de Ciencias Exactas e Ingenierías, Universidad de Guadalajara, Guadalajara, Jalisco, México

2 Departamento de Ingeniería de Proyectos, Centro Universitario de Ciencias Exactas e Ingenierías, Universidad de Guadalajara, Guadalajara, Jalisco, México

3 Departamento de Ciencias Básicas, Centro Universitario de la Ciénega, Universidad de Guadalajara, Ocotlán, Jalisco, México

4 Departamento de Ciencias Computacionales e Ingenierías, Centro Universitario de los Valles, Universidad de Guadalajara, Ameca, Jalisco, México

5 Departamento de Química, Centro Universitario de Ciencias Exactas e Ingenierías, Universidad de Guadalajara, Guadalajara, Jalisco, México

6 Departamento de Ciencias Tecnológicas, Centro Universitario de la Ciénega, Universidad de Guadalajara, Ocotlán, Jalisco, México

*Address all correspondence to: trinidad.guillen@academicos.udg.mx

References

[1] Schütze A, Baur T, Leidinger M, Reimringer W, Jung R, Conrad T, et al. Highly sensitive and selective VOC sensor systems based on semiconductor gas sensors: How to? Environments. 2017;**4**:20. DOI: 10.3390/environments 4010020

[2] Nandy T, Coutu RA, Ababei C. Carbon monoxide sensing Technologies for Next-Generation Cyber-Physical Systems. Sensors. 2018;**18**:3443. DOI: 10.3390/s18103443

[3] Wimmer-Teubenbacher R, Sosada-Ludwikowska F, Zaragoza TB, Defregger S, Tokmak O, Steffen NJ, et al. CuO thin films functionalized with gold nanoparticles for Conductometric carbon dioxide gas sensing. Chemical Sensors. 2018;**6**:56. DOI: 10.3390/chemosensors6040056

[4] Wang C, Yin L, Zhang L, Xiang D, Gao R. Metal oxide gas sensors: Sensitivity and influencing factors. Sensors. 2010;**10**:2088-2106. DOI: 10.3390/s100302088

[5] Comini E, Baratto C, Concina I, Faglia G, Falasconi M, Ferroni M, et al. Metal oxide nanoscience and nanotechnology for chemical sensors. Sensors and Actuators B. 2013;**179**:3-20. DOI: 10.1016/j.snb.2012.10.027

[6] Eranna G, Joshi BC, Runthala DP, Gupta RP. Oxide materials for development of integrated gas sensors—A comprehensive review. Critical Reviews in Solid State and Materials Sciences. 2004;**29**:111-188. DOI: 10.1080/10408430490888977

[7] Addabbo T, Bertocci F, Fort A, Gregorkiewitz M, Mugnaini M, Spinicci R, et al. Gas sensing properties and modeling of YCoO3 based perovskite materials. Sensors and Actuators B. 2015;**221**:1137-1155. DOI: 10.1016/j.snb.2015.07.079

[8] Gildo-Ortiz L.; Reyes-Gómez J.; Flores-Álvarez J.M.; Guillén-Bonilla H.; de la Olvera M. L. de la L.; Rodríguez Betancourtt V.M.; Verde-Gómez Y.; Guillén-Cervantes A.; Santoyo-Salazar J. Synthesis, characterization and sensitivity tests of perovskite-type LaFeO3 nanoparticles in CO and propane atmospheres. Ceramics International 2016, 42, 18821–18827. DOI:10.1016/j.ceramint.2016.09.027

[9] Yuan Y, Wang B, Wang C, Li X, Huang J, Zhang H, et al. Effects of CoFe2O4 electrode microstructure on the sensing properties for mixed potential NH3 sensor. Sensors and Actuators B. 2017;**239**:462-466. DOI: 10.1016/j.snb.2016.07.171

[10] Dong C, Liu X, Xiao X, Du S, Wang Y. Monodisperse ZnFe2O4 nanospheres synthesized by a nonaqueous route for a highly selective low-ppm-level toluene gas sensor. Sensors and Actuators B. 2017;**239**: 1231-1236. DOI: 10.1016/j.snb.2016.09.122

[11] Singh S, Singh A, Singh A, Rathore S, Yadav BC, Tandon P. Nanostructured cobalt antimonate: A fast responsive and highly stable sensing material for liquefied petroleum gas detection at room temperature. RSC Advances. 2020; **10**:33770-33781. DOI: 10.1039/D0RA06208A

[12] Singh A, Singh A, Singh S, Tandon P. Nickel antimony oxide (NiSb2O6): A fascinating nanostructured material for gas sensing application. Chemical Physics Letters. 2016;**646**:41-46. DOI: 10.1016/j.cplett.2016.01.005

[13] Guillén-Bonilla H, Flores-Martínez M, Rodríguez-Betancourtt VM, Guillén-Bonilla A, Reyes-Gómez J, Gildo-Ortiz L, et al. A novel gas sensor based on $MgSb_2O_6$ Nanorods to indicate variations in carbon monoxide and propane concentrations. Sensors. 2016; **16**:177-188. DOI: 10.3390/s16020177

[14] Singh S, Singh A, Singh A, Tandon P. A stable and highly sensitive room-temperature liquefied petroleum gas sensor based on nanocubes/cuboids of zinc antimonate. RSC Advances. 2020; **10**:20349-22357. DOI: 10.1039/D0RA02125C

[15] Rodríguez-Betancourtt VM, Guillén BH, Flores MM, Guillén BA, Moran Lazaro JP, Guillén Bonilla JT, et al. de la L. gas sensing properties of NiSb2O6 micro- and nanoparticles in propane and carbon monoxide atmospheres. Journal of Nanomaterials. 2017;**2017**:1-9. DOI: 10.1155/2017/8792567

[16] Haeuseler H. Infrared and Raman spectra and normal coordinate calculations on trirutile-type compounds. Spectrochimica Acta. 1981; **37A**:487-495. DOI: 10.1016/0584-8539(81)80036-0

[17] Giere EO, Brahimi A, Deiseroth HJ, Reinen D. The geometry and electronic structure of the Cu^{2+} Polyhedra in Trirutile-type compounds Zn (Mg)1-xCuxSb_2O_6 and the dimorphism of $CuSb_2O_6$: A solid state and EPR study. Journal of Solid State Chemistry. 1997; **131**:263-274. DOI: 10.1006/jssc.1997.7374

[18] Reimers JN, Greedan JE. Crystal structure and magnetism in MnSb2O6: Incommensurate long-range order. Journal of Solid State Chemistry. 1989; **79**:263-276. DOI: 10.1016/0022-4596(89)90273-9

[19] Kinoshita M, Seki S, Sato TJ, Nambu Y, Hong T, Matsuda M, et al. Magnetic reversal of electric polarization with fixed chirality of magnetic structure in a chiral-lattice Helimagnet $MnSb_2O_6$. Physical Review Letters. 2016; **117**:047201. DOI: 10.1103/PhysRevLett.117.047201

[20] Guillén-Bonilla H, Rodríguez-Betancourtt VM, Guillén-Bonilla JT, Gildo-Ortiz L, Guillén-Bonilla A, Casallas-Moreno YL, et al. Sensitivity tests of pellets made from manganese Antimonate nanoparticles in carbon monoxide and propane atmospheres. Sensors. 2018;**18**: 2299. DOI: 10.3390/s18072299

[21] Casillas Zamora A, Guillen Bonilla JT, Guillen Bonilla A, Rodríguez Betancourtt M, Casallas Moreno YL, Gildo Ortiz L, et al. Synthesis of $MnSb_2O_6$ powders through a simple low-temperature method and their test as a gas sensor. Journal of materials science: Materials. Electronics. 2020;**31**: 7359-7372. DOI: 10.1007/s10854-019-02700-3

[22] Scott HG. Synthesis and crystal structures of the manganous antimonates $Mn_2Sb_2O_7$ and MnSb2O6. Journal of Solid State Chemistry. 1987; **66**:171-180

[23] Singh J, Bhardwaj N, Uma S. Single step hydrothermal based synthesis of M(II)Sb_2O_6 (M = Cd and Zn) type antimonates and their photocatalytic properties. Bulletin of Materials Science. 2013;**36**:287-291

[24] Mizoguchi H, Woodward PM. Electronic structure studies of main group oxides possessing edge-sharing octahedra: Implications for the design of transparent conducting oxides. Chemistry of Materials. 2004;**16**: 5233-5248

[25] Delgado E, Michel CR. CO_2 and O_2 sensing behavior of nanostructured

barium-doped SmCoO3. Materials Letters. 2006;**60**:1613-1616

[26] Rodríguez Betancourtt V-M, Guillén-Bonilla H, Guillén-Bonilla JT, Casallas-Moreno YL, Ramírez-Ortega JA, Morán-Lázaro JP, et al. Synthesis, characterization, and sensitivity tests of a novel sensor based on barium antimonate powders. Materials Today Communications. 2022;**31**: 103579. DOI: 10.1016/j.mtcomm. 2022.103579

[27] Nalbandyan VB, Zvereva EA, Nikulin AY, Shukaev IL, Whangbo M, Koo H, et al. New phase of MnSb2O6 prepared by ion exchange: Structural, magnetic, and thermodynamic properties. Inorganic Chemistry. 2015; **54**:1705-1711

[28] Gorbunov VV, Shidlovskii AA, Shmagin LF. Combustion of transition-metal ethylenediamine nitrates. Combustion, Explosion, and Shock Waves. 1983;**19**(2):172-173

[29] Wang X, Li Y. Solution-based synthetic strategies for 1-D nanostructures. Inorganic Chemistry. 2006;**45**:7522-7534

[30] Deng Z-X, Wang C, Sun X-M, Li YD. Structure-directing coordination template effect of ethylenediamine in formations of ZnS and ZnSe nanocrystallites via solvothermal route. Inorganic Chemistry. 2002;**41**:869-873

[31] Michel CR, Martínez-Preciado AH, Morán-Lázaro JP. Effect of the frequency on the gas sensing response of $CoSb_2O_6$ prepared by a colloidal method. Sensors and Actuators B. 2009;**140**: 149-154

[32] Bochenkov VE, Sergeev GB. Preparation and chemiresistive properties of nanostructured materials. Advances in Colloid and Interface Science. 2005;**116**(1–3):245-254

[33] Jayaraman VK, Álvarez AM, de la Luz Olvera Amador M. A simple and cost-effective zinc oxide thin film sensor for propane gas detection. Materials Letters. 2015;**157**:169-171

Chapter 5

Numerical Analysis of the Steady State in SAW Sensor Structures with Selected Polymers for Detection of DMMP and CO

Tomasz Hejczyk, Jarosław Wrotniak and Wiesław Jakubik

Abstract

The chapter presents the results of the numerical investigation of the SAW gas detector structures with selected polymer layers in steady-state conditions. The effect of SAW velocity changes vs. the surface electrical conductivity of the detector structures is predicted on the base of acoustoelectric elemental theory. The electrical surface conductivity of the rough polymer sensing layer placed above the piezoelectric waveguide depends on the profile of the diffused gas molecule concentration inside the whole detector structure. Numerical results in the steady state conditions for the gas molecules DMMP and polymer layer of (RR)-P3HT have been shown as well as for carbon oxide molecules with thin polyaniline and polypyrrole layer. The main aim of the investigations was to study a thin film's interaction with targeted gases in the SAW detector configuration based on diffusion equations for polymers. Numerical results for profile concentration in steady state conditions for gas molecules concentration, film thickness, roughness, and interaction temperature have been shown. The results of numerical analyzes allow for selecting better detector design conditions, including the morphology of the detector layer, its thickness, operating temperature, and layer type. The numerical results, based on the code written in Python, were shown.

Keywords: SAW (surface acoustic wave) gas detector, SAW, DMMP (dimethyl methylphosphonate) detection, polymer, (Regio-regular)-P3HT, (RR)-P3HT, polyaniline (PANI), polipirol (PPy), Nafion, numerical acoustoelectric analysis (NAA), Ingebrigtsen's formula, CO (carbon monoxide), python

1. Introduction

Polymers are a very interesting material for testing the concentration of particularly dangerous gases, CO (the silent killer), and are also used in research on toxic warfare agents. The following publication presents two approaches that have evolved during research, both in terms of the laboratory method—optimization of the measurement system in terms of control, and thanks to technological progress and the use of atomic force microscopy (AFM). This lastly allows for the emergence of the

IntechOpen

polymer structure and confirmation of the dependence of the shape of the detector surfaces, its morphology, and its influence on the SAW detector response. Historically, different thin film materials and the piezoelectric substrates ($LiNbO_3$—lithium niobate Y cut-Z propagation) were utilized with the detector layers ranging from semiconducting to polymers—on which we focused in our research work—PANI (polyaniline), PPY (polypyrrole) with NAFION—to detect carbon oxide (CO), and RR-P3HT (regio-regular poly-3-hexylothiophen)—to detect DMMP (dimethylo-methylo-phosphonate) [1]—a simulant of the poison of chemical warfare agent (CWA), like sarin. The development and optimization of the NNA direction (normalized numerical acoustoelectric analysis) over the analytical detector with an acoustic surface wave (SAW), which now, thanks to technological development, has been empirically confirmed.

Over the decade, the research team made efforts to select the appropriate detector layer, its thickness, porosity, and morphology for individual gases that are important for human safety, e.g. in CO detectors [2]. This publication was created from these research works. Particular importance, also from the point of view of safety, are polymer layers, those photoconductive, which are used to detect CWA (Chemical Warfare Agents), the use of which in systems can significantly prevent a tragedy. For the purposes of the experiment, an original analytical model of the gas detector was created. Initially, this model was based on Knudsen diffusion mechanisms and was inspired by the works of: Matsumaga N., Sakai G., Shimanoe K., Yamazoe N [3, 4]. The main aim of the investigation was to study thin film interaction with target gases in the SAW detector based on a simple reaction-diffusion eq. [5–11]. Diffusion equations provide theoretical bases for the analysis of physical phenomena like heat transport or mass transport in porous, roughness substrates [5].

In the initial version of the development of the analytical model, the porosity of the layer was taken into consideration. This model was a good illustration of the operation of semiconductor gas detectors. However, in the case of polymer layers and due to different physicochemical properties [12] of polymer layers and other differential equations, this model had to evolve toward a model that took into account the roughness of the layer.

Generally, these detectors have a very high sensitivity, much higher than the commercially available resistance gas detector. Detectors based on surface acoustic waves are widely used in many industries, especially in biochemical applications, allowing for monitoring of DNA mutation [13] and commercial applications, such as monitoring the quality of food, as well as monitoring the physical and chemical properties of solids, such as adsorption/desorption of the substance, humidity.

The chapter summarizes the acoustoelectric theory, i.e. Ingebrigtsen's formula, dynamics gas diffusion concentration profiles in steady-state, and predicts the influence of a thin polymer detector layer with new polymer gas diffusion model on the SAW wave velocity in a piezoelectric acoustic waveguide [14].

Delay lines, with acoustic surface wave (SAW), enable the detection of very small concentrations of chemical compounds in gas mixtures [15]. The miniaturization of these transducers resulted in a significant increase in the frequency of SAW detector. However, miniaturization of the detector requires the construction of more complicated or technically advanced electronic devices (**Figure 1**), while the test and measurement system for gas detection [6, 17–20] has been designed and developed independently.

The detection materials used here, such as DMMP, are not only valuable due to the fact that they can be used as sarin simulants for the calibration of organophosphorus

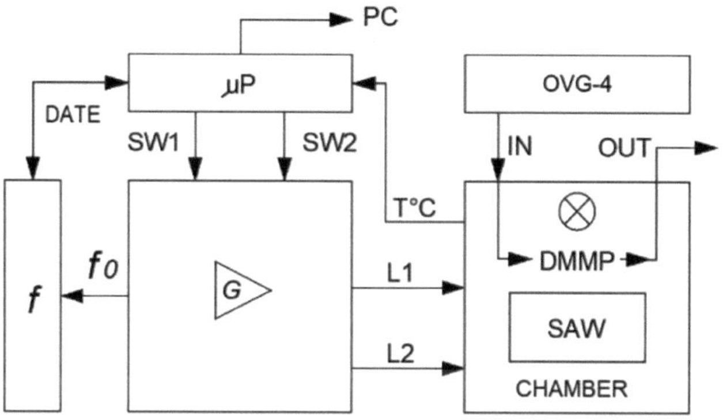

Figure 1.
Measurement appliance—patent no PAT.230526—system for detecting chemical compounds in gaseous atmospheres, with a detector using surface acoustic waves (SAW) and digital switches SW1 and SW2 ([16], µP—microprocessor, OVG-4—the owlstone calibration gas generator, G—generator design and implementation by authors: J. Wrotniak and M. Magnuski).

detectors, but also due to the obvious fact that DMMP can be used in the production of chemical weapons (sarin and soman) and is applicable to the construction of very sensitive detectors of this unnoticeable lethal weapon—chemical weapon, which in war conditions (in violation of international rules of war) has a special military and strategic importance.

Chemical warfare agents (CWA), especially nerve agents (e.g. sarin and soman) are highly lethal compounds. Sarin is one of the best-known chemical weapons of mass destruction. This odorless and colorless compound causes neuromuscular paralysis and death by suffocation within 1–10 min. Disabling and lethal exposures to sarin occur above 15 ppb and 64 ppb, respectively for 10 min of exposure [21].

In gas detectors based on SAW, the mechanism of detecting the concentration of the gas or vapors of a chemical compound depends on the interaction of its molecules with a properly selected detector layer sensitive to its presence [22]. The processes of interaction between gas molecules with the layer are kinetic phenomena, mainly sorption (in volume) and adsorption (on the surface), resulting from the entrapment of the molecule in the layer or on its surface. Sorption of gas or vapor molecules through the detector layer causes a change in its mass and electrical conductivity (change in conductivity affects the change of SAW propagation velocity) which in the measurement system leads to a change in the generator's operating frequency [23]. The channel with the detector layer generates as a result oscillations with a different frequency (usually lower) and is shifted in phase relative to the signal generated in the reference path (**Figure 2a**). The work focused on the electrical effect is a new contribution to the SAW gas detector technology [7].

The results of the research on the application of RR-P3HT (poly 3-hexylthiophene regioregular type) produced by means of air spraying on a quartz module with SAW 205 MHz to detect traces of DMMP (Dimethyl methylphosphonate) molecules were presented [1, 15]. DMMP is a non-toxic substance with a similar chemical structure to sarin (Combat Poisoning Agent) and it allows for safe experiments.

Due to the photoconductive properties of the P3HT polymer [24], the layer was additionally activated by LED light (**Figure 2b**). The sensitivity of the layer in the system with SAW [25] to the presence of DMMP in this manner was increased (different wavelengths of LED illumination). Oscillations (in reference and measurement line) were excited. Generator 205 MHz with switched channels was used.

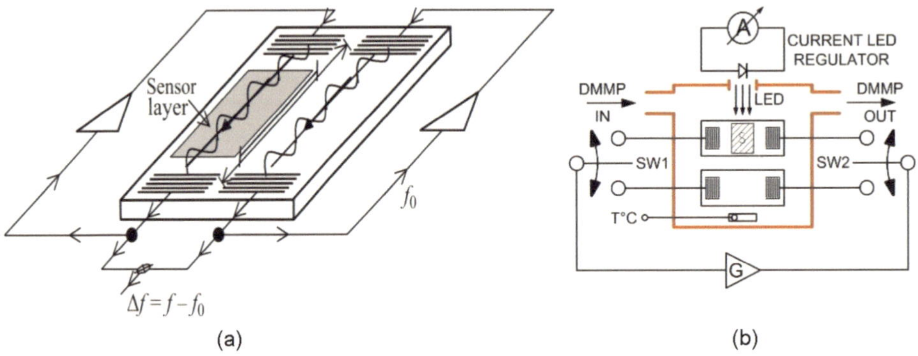

Figure 2.
Measuring appliance: (a) the idea of the historical measuring system [2, 14], and (b) the newest measurement system with LED lighting with switched channels [1, 8, 16].

The essence of the conducted research is the search for new materials and ways to activate them in order to detect trace concentrations of DMMP in the air without the need of applying high temperatures (above 100°C) [1].

2. Technology of applying the detector layer PANI and (RR)—P3HT and sensing mechanism

The bilayer structure of Nafion-Polyanliline was examined by the acousto-electric method (**Figure 2a**). An electrometer Keithley was used in the study due to the high resistance of the structure. The main purpose of the Nafion application was the possibility of controlled protonation of the second layer based on polyaniline to increase its electrical conductivity. Nafion (thickness of approx. 300 nm) and polyaniline (for the thickness of the layers: 100, and 180 nm) were used as sensing structures [2]. A thin layer of Nafion was deposited on the surface of the $LiNbO_3$ waveguide by the spin-coating method at speed of 7500 rpm. Thin layers of Nafion were annealed at 40°C for 2 hours and then for 15 minutes at 120°C to harden them. Polyaniline (PANI) layers of thicknesses of 100 nm and 180 nm were made in a vacuum evaporation process. The whole process was carried out in the residua atmosphere of argon (Ar), in connection with the oxidation of the polymer during the time of its application, which affected the sensing properties of the layers. To achieve the desired quality of the atmosphere in the vacuum chamber, the set-up was rinsed three times with argon. Then, in order to evaporate the water vapor trapped in the PANI, the structure was annealed at 200°C for 10 minutes. The polyaniline application process was made at a temperature of 350–400°C. To obtain the desired thickness, the sublimation process lasted about 45 minutes.

The detector layer (**Figure 3**) of the (RR)—P3HT type polymer with a thickness of ~350 nm, in the empty space of one of the delay lines of the quartz module with SAW was created. The thickness using the atomic force microscope (AFM) profile analysis was estimated. A field fragment polymer plate to use a suitably designed mask was exposed. Through the open window spraying method (nozzle thickness—0.4 mm) a pre-prepared polymer solution (RR)—P3HT was applied. Compressed synthetic air at a pressure of 1 atm. Was used. Solution by dissolving about 1 mg (RR)—P3HT in 1 ml chloroform was prepared. The distance of the nozzle from the substrate during the process was: about 40 mm, settling time was about 3 s (see **Figure 4**) [1, 8].

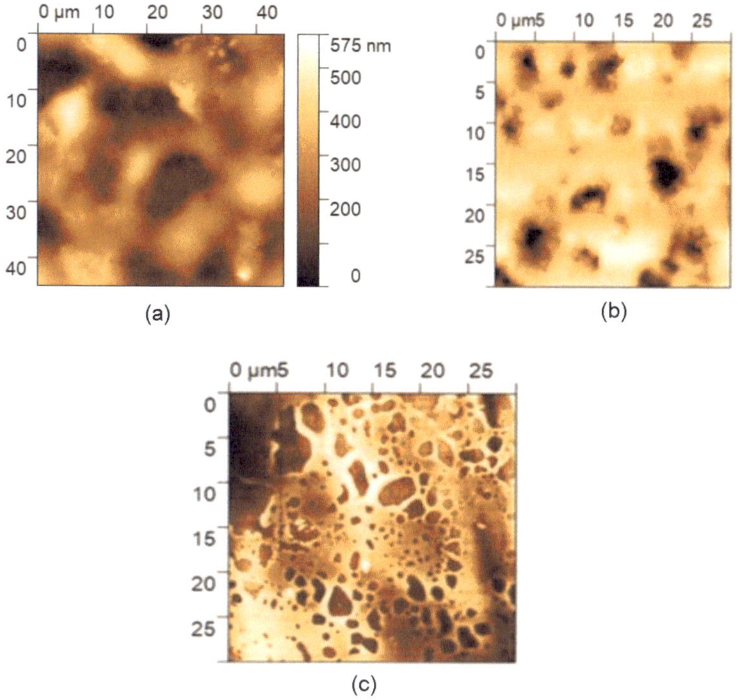

Figure 3.
Layer topography/morphology (RR) P3HT quartz substrate—Fragment of polymer layer (RR)-P3HT on quartz crystal from SAW—View under magnification (a), AFM images of the P3HT-RR sensing layer measured on a laminate: The polymer surface—resistive detectors research (b) AFM images of the P3HT-RR sensing layer on metallization: The polymer surface—Resistive detectors research (c) [1].

Figure 4.
View of a fragment of the applied polymer layer (RR)-P3HT on the quartz crystal of the AFP module. View of the boundary of the formation of a porous layer from the crystal edge visible from above 100x magnification [1].

The sensing mechanism in PANI or PPY (Polipyrrole) and P3HT detectors layers is very interesting. The interaction of CO molecules with PANI is primarily of electrical nature. CO is an oxidizing gas—CO molecules attach electrons from the PANI structure. Polyaniline is a p-type semiconductor. The binding of PANI electrons to CO molecules results in an increase in the electrical resistance of the structure (the difference between the concentration of holes and electrons increases). This is manifested by a decrease in the frequency of the measuring resonator and, as a result, a decrease in Δf relative to the situation without CO particles in the atmosphere. In the measurements, the sensing structure was tested for the presence of very low concentrations of CO (5, 10, 15, and 20 ppm) in the synthetic air.

The P3HT contains delocalized π bonds, which permit the easy flow of electrons within the delocalized π system. The relative response of P3HT to analyte species depends on their Lewis acidity or basicity: stronger acids or bases have a larger effect on polymer resistance, with acids decreasing resistance and bases increasing resistance. For DMMP, there is a phosphorous-oxygen double bond. The electronegativity of the oxygen atom is stronger than that of the phosphorus atom, resulting in the increase of the electron density of the oxygen, so DMMP shows alkalescence. The p-type semiconducting behavior of P3HT promotes holes in the valence band of P3HT which play a key role in sensing properties. The DMMP is a strong electron donor which depletes holes from the valence band of P3HT, resulting in an increase of resistance after being adsorbed on the P3HT surface. The number of charge carriers in the P3HT film is increased by the light excitation, resulting in enhanced sensing properties, like sensitivity, the limit of detection, and response time. In addition, for the highly developed surface of the sensing layer deposited on the porous substrate, we obtain more active adsorption sites, and the scheme of the sensing mechanism described above is presented in [15].

3. Experimental investigations

In the research, a measuring stand with resonators on acoustic surface waves SAW with a positive feedback loop was used in historical measurements (**Figure 2a**). The system consists of two identical delay lines—DLs (or resonators). One of the delay lines (or resonator) is isolated from the influence of the external atmosphere. The second line (resonator) is exposed to an external gas environment. Changes in the chemical composition of the atmosphere change the resonance frequency of the active line (resonator). At the output of the set-up (DLs or resonators), their high-frequency signals are electronically mixed. The frequency of the delay line without the structure of the detectors was 43.60 MHz, while the sensing structure was lower from several dozen to even one hundred kHz (as a result of its mass loading by the sensing structure). The normal mode frequency configuration (NMFC) occurs when the measuring frequency (f) is lower than the reference ones (f_0). In the investigations as mentioned above, the bilayer sensing structures of Nafion+Polyaniline were examined by means of the acoustoelectric method in the NMFC case and the difference frequency Δf is determined as $f–f_0$ (see **Figures 5** and **6**).

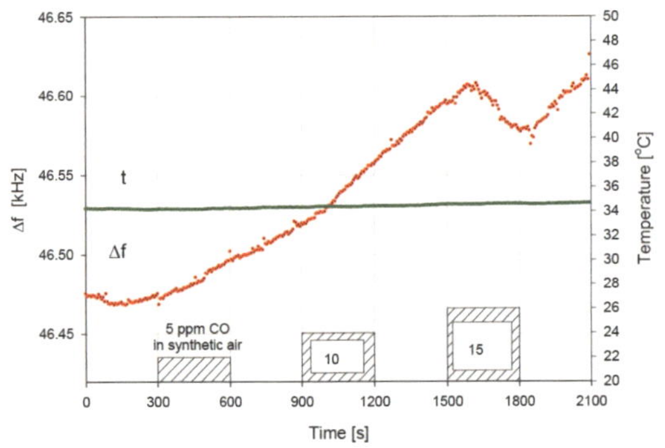

Figure 5.
Response (Δf) of the bilayer detector structure PANI (100 nm) + Nafion, to CO gas (5, 10, 15 ppm) in air, T = 34°C.

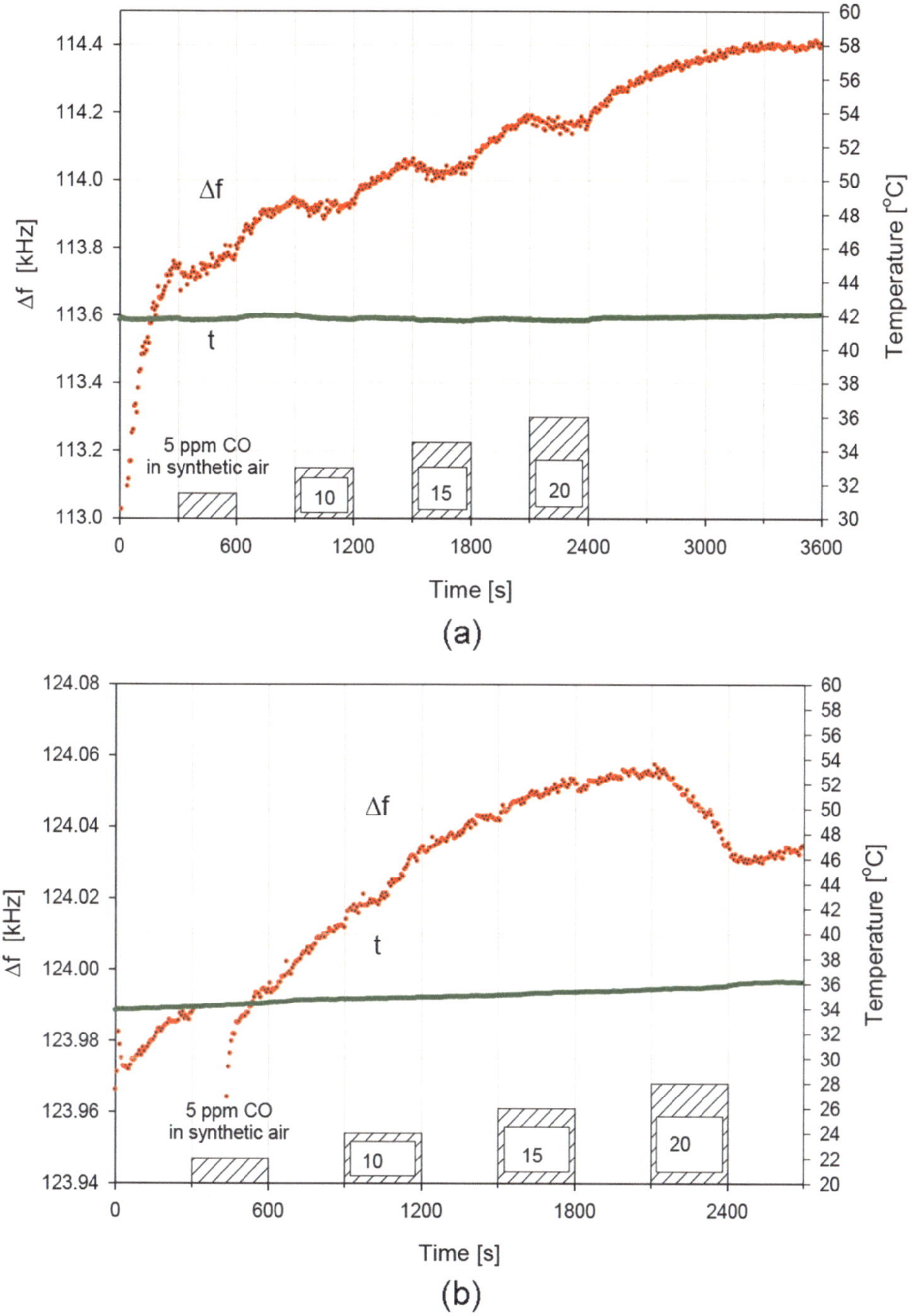

Figure 6.
Response (Δf) of the bilayer detector structure PANI (180 nm) + Nafion, and CO gas (5, 10, 15, 20 ppm), T = ~42° C (a), response (Δf) of the detector layer PANI (180 nm) + Nafion, CO gas (5, 10, 15, 20 ppm), T = ~35°C (b) [2].

The results of the studies prepared for the numerical acoustoelectric analysis (NNA) studies for the PANI layer below were shown.

The collective responses (Δfmax-Δf)/Δfmax) SAW detector on CO gas at thicknesses 100 nm and 180 nm (PANI), for the concentration from 5 to 15 ppm in synthetic air at a temperature of 34° C (100 nm) and 35° C (180 nm) in **Figure** 7 were showed (**Figure 8**) [2].

The work presents comparative measurements for a polymer layer based on Polypyrrole. The results of the experiment below were shown (**Figure 9**).

Comparison of the detection properties of Polyaniline + Nafion layers (film thickness of 100 nm and 180 nm) and Polipirol layer (thickness 80 nm) of the concentration of 25 ppm, 37.5 ppm, 50 ppm in **Figure 10** were shown. It is clear that the Polypyrrole layers will be suitable for the measurement of higher concentrations.

The main aim of researching selected detector structures with a surface acoustic wave was an empirical verification of the response with numerical analysis. We must emphasize that performing empirical research was possible within limits because of

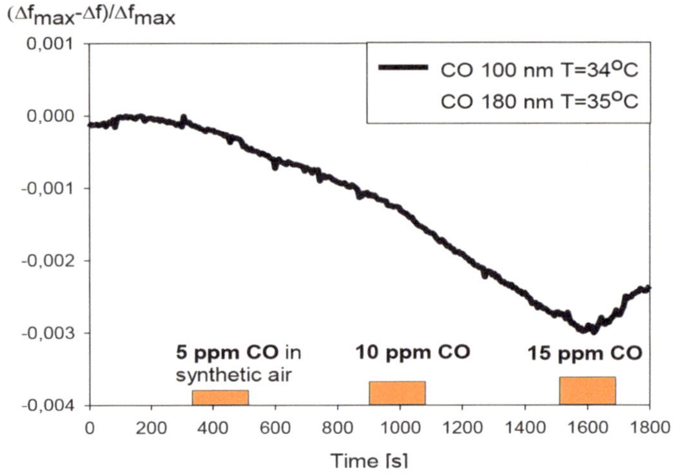

Figure 7.
Relative changes of the response of a detector normed to maximum differential frequency for each layer PANI +Nafion: 100, 180 nm, CO gas (5, 10, 15 ppm), in synthetic air—measuring system ***Figure 2a.***

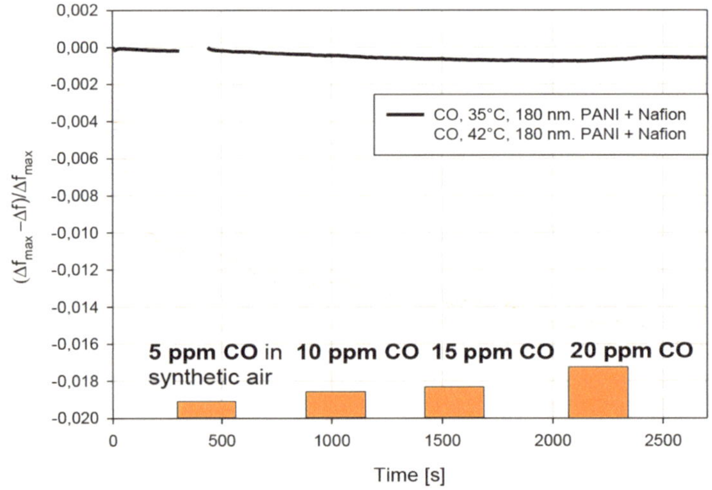

Figure 8.
Relative changes of the response of a detector normed to maximum differential frequency for PANI+Nafion 180 nm, CO gas (5, 10, 15, 20 ppm), at temperature: 35°C and 42°C—measuring system ***Figure 2a.***

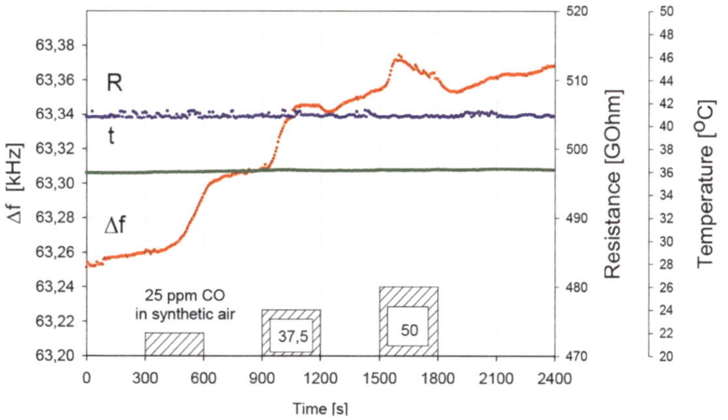

Figure 9.
Acoustical response (Δf) and resistance response (R) of the detector layer PPy (Polypyrrole 80 nm), CO gas (25, 37.5, 50 ppm), T = 36°C—measuring system ***Figure 2a.***

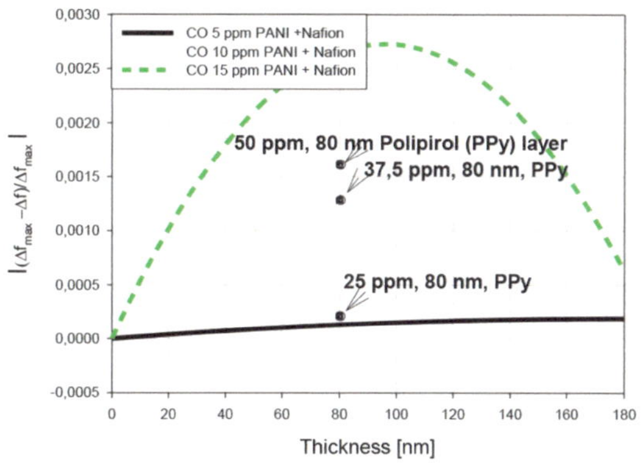

Figure 10.
A comparison of the detection properties of PANI + NAFION with empirical results of Polypyrrole layer (80 nm) at temperature = 35°C—measuring system ***Figure 2a.***

the wide range of work and the complex technological processes connected with the practical feasibility of the detector structure [1, 16].

A practical system for testing the acoustic-electrical properties of polymer layers based on (Regio-Regular)-P3HT in **Figure 11** was shown.

Empirical results for illuminating LEDs (200 mA) with different wavelengthsof the layer (RR)-P3HT are shown in **Figure 12**. The research system from **Figure 2b** was used. The measuring system uses diodes—1 W SMD 350 mA ProLight Opto.

Figure 11.
The practical implementation of the invention of the patent no. PL 230526 B1—A measuring chamber for testing using SAW detectors—measuring system ***Figure 2b.***

Results of measurements DMMP in interaction with polymer layer RR-P3HT illuminated with a diode light of 200 mA in the form of a histography are shown in **Figure 13**.

In order to optimize numerically and compare the experiment with the theory, the results of the experiment depending on the concentration were normalized as follows in **Figure 14**.

Three measurement series were made. The diodes were driven with 100, 200, 300 mA current. The layer was additionally activated by a small incandescent lamp with white light (with a maximum of about 750 nm) and different illuminance (see **Figure 13**). Exposure time was about 10 min for each 100, 200, and 300 mA current.

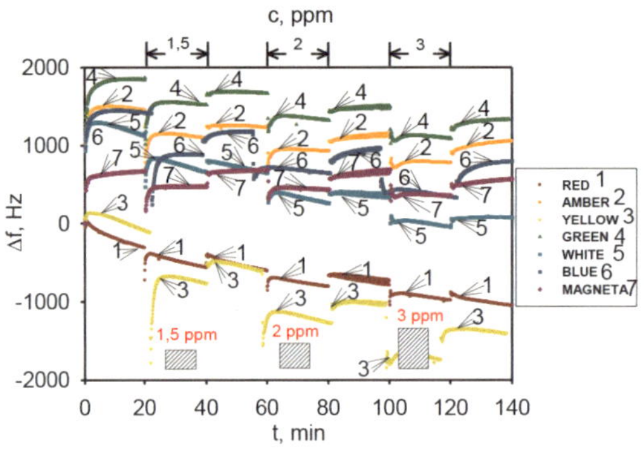

Figure 12.
Experiment—Detector layer (RR)-P3HT, thickness 500 nm, gas DMMP (1.5; 2; 3 ppm, illumination by diode 200 mA (selected wavelengths) relative change of velocity vs. time (concentration)—measuring system ***Figure 2b****.*

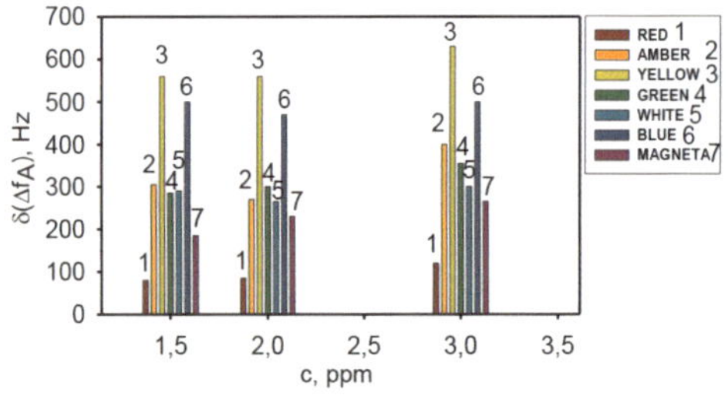

Figure 13.
Experiment—Layer (RR)-P3HT, thickness 500 nm, gas DMMP (1.5; 2; 3 ppm, illumination by diode 200 mA (selected wavelengths)—Histography—relative change of velocity vs. time (concentration)—measuring system ***Figure 2b****.*

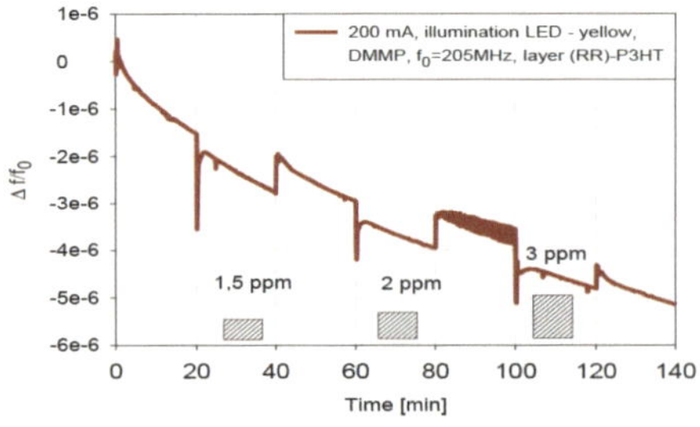

Figure 14.
Experiment—Layer (RR)-P3HT, thickness 500 nm, gas DMMP (1.5; 2; 3 ppm, illumination by diode 200 mA (yellow light) normalized results relative change of velocity vs. time (concentration)—measuring system ***Figure 2b****.*

In order to optimize numerically and compare the experiment with the theory, the results of the experiment were normalized as follows in **Figure 14**—depending on the concentration. Results with a DMMP concentration of 2 ppm in the air were shown.

4. Numerical analysis (NNA) of the acoustoelectric interaction in the sensing layer in the steady state conditions

The acoustic-electric effect depends on the electrical charge profile distribution in the sensing layer on the absorbed gas particles' distance from the surface acoustic waveguide.

To determine the response of the detectors common impedance was designated. Impedance includes information about the profile of the concentration of gas molecules in the layer and has been implemented into the Ingebrigtsen formula [23]. This enables describe relative change velocity surface acoustic waves in steady state and transient mode. Analytical expressions clearly define the model of the SAW detectors. Based on this model numerical analysis detector's response was made. The results of numerical analyzes are shown in the next section.

The analytical model of a SAW gas detector having the described form was used in order to analyze it numerically. In order to analyze such a detector layer in the SAW gas detector we assumed that the film is a uniform stack of infinitesimally thin sheets with a variable concentration of gas molecules (**Figure 15**) and that it influences the electrical conductance [10].

$$\frac{\Delta v}{v_0} = -\operatorname{Re}\left\{\frac{\Delta k}{k_0}\right\} = -\frac{K^2}{2}\frac{\left[\sigma_{T_2}\left(1+aC_{A,y=0}\right)+\sum_{i=1}^{n-1}\sigma_{T_2}(y_i)f\left(y_i,\sigma_{T_2}(y_i)\right)\right]^2}{\left[\sigma_{T_2}\left(1+aC_{A,y=0}\right)+\sum_{i=1}^{n-1}\sigma_{T_2}(y_i)f\left(y_i,\sigma_{T_2}(y_i)\right)\right]^2+\left[1+\sum_{i=1}^{n-1}g\left(y_i,\sigma_{T_2}(y_i)\right)\right]^2(v_0F_S)^2} \quad (1)$$

where: n—number of sublayers and $F_S = \varepsilon_0 + \varepsilon_p^T$, i = 1, 2, 3, ... n (the sublayers index), $\Delta v/v_0$ and $\Delta k/k_0$ relative changes of velocity and wave vector of SAW,

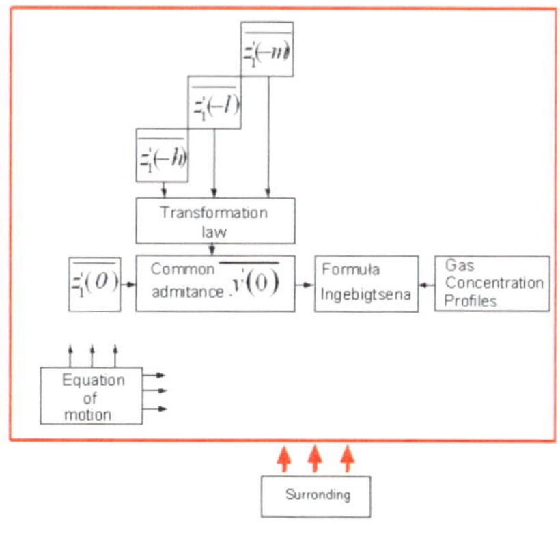

Figure 15.
Schematic diagram of the SAW detector model [2].

respectively, σ_0—electrical conductivity of sensing layer in air, K^2—coefficient of electromechanical coupling,

$$\sigma(y_i) = \sigma_0 \left[1 + a \cdot C_A(y_i)\right] \tag{2}$$

where: *a*—sensitivity coefficient of the sensing layer [1/ppm].

In the expression (1) the functions $f(y_i, \sigma(y_i))$ and $g(y_i, \sigma(y_i))$ are the results of the transformation of the individual sublayer on the surface of a detector waveguide (**Figure 15**) and it has the form:

$$f(y_i, \sigma(y_i)) = \frac{1 - [\mathrm{tgh}(ky_i)]^2}{[1 + \mathrm{tgh}(ky_i)]^2 + \left[\mathrm{tgh}(ky_i) \cdot \frac{\sigma(y_i)}{\varepsilon_0 v_0}\right]^2} \tag{3}$$

and

$$f(y_i, \sigma(y_i)) = \frac{1 - [\mathrm{tgh}(ky_i)]^2}{[1 + \mathrm{tgh}(ky_i)]^2 + \left[\mathrm{tgh}(ky_i) \cdot \frac{\sigma(y_i)}{\varepsilon_0 v_0}\right]^2} \tag{4}$$

The conductivity of the detector layer depends on the temperature:

$$\sigma_{T_2} = \sigma_{T_1} \exp\left(\frac{E_g}{2k_B} \cdot \frac{T_2 - T_1}{T_1 T_2}\right) \tag{5}$$

where $T_1 = 300$ K, $\sigma_{T_1} = \sigma_0$ are the conductance of the layer, respectively, at temperature the T_1 and T_2, k_B—the Boltzmann constant, v_o—SAW velocity, k—acoustic wave number ($k = 2\pi/\lambda$), E_g—the width of the energy gap of detector layer material. The expressions (1, 3, 4) make it possible to determine the use of the iteration method response of the surface acoustic wave in the steady-state ($t \rightarrow \infty$).

For the selected parameters (thickness, concentration, temperature, sensitivity *a*, and conductivity) of bilayer sensor structure (Nafion+Polyanyline), the numerical analysis were made and the influence of carbon monoxide gas concentration on detection response was determined.

Figure 16 presents the numerical analysis detector response depending on the thickness detector layer. From the analysis, it follows that the optimum thickness is approx. 90 nm. The software was written in Python. In the range of thicker layers

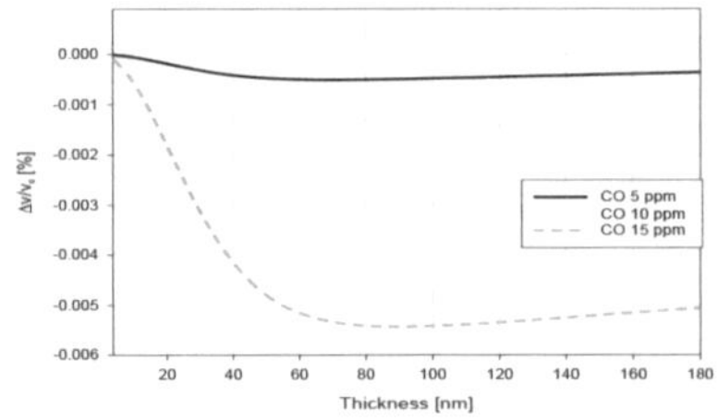

Figure 16.
Changes of velocity (relative) vs. thickness from 2 nm to 180 nm, Nafion+Polyanyline. Assumed in the analysis: Sensitivity coefficient a = −4.25 ppm^{-1}*,* $\sigma_s = 2 \times 10^{-12}$ *S, M = 28 g/Mol (CO), concentration: 5, 10, 15 ppm. Numerical results.*

above 180 nm interaction decreases, which confirms obtained empirical results. From the characteristic, it follows that for the small concentration (below 5 ppm) measurements of carbon monoxide using an SAW detector for the layer above 180 nm and thicker will be useless (measurement temperature of 35°C, see **Figures 5, 6a**). This fact very explicitly confirms the experiment mentioned above for thickness layer of 180 nm and at the temperature of 35°C.

The numerical analysis of the influence of the CO concentration on the response of the detector in a layer thickness of 180 nm for a temperature of 35° C and 42° C (**Figure 6b**) was made (**Figure 17**).

A separate place is devoted to researching the temperature properties of the layer. It is important here to note that the selection of the operating temperature of the detector depends not only on the type of detector layer but also on the porosity and roughness (**Figure 18a** and **b**). Temperature change also allows to determine the operating point of the detector and also has an impact on the speed of response and recovery of the detector layer. Changing the operating point of the detector by increasing the temperature from 35°C to 42° C, should increase the detection properties of the detector in accordance with the experiment (**Figure 6a** and **b**). The effect of temperature on the response of the gas detector is also confirmed by the analysis of the properties of the temperature detector (**Figure 8**). The numerical analysis of the influence of the temperature on the response of the detector in a layer thickness of 180 nm for a temperature of 35°C (308 K) and 42°C (315 K) was made (**Figure 19**). Numerical analysis shows that based on the numerical model of a gas detector with a layer of Nafion + Polyaniline gas impact in the case of measuring the temperature increase of 35–42°C will result in a marked increase in the detector response.

The results of numerical analysis (**Figure 19**) coincide approximately with the results of the experiment (see **Figures 5–8**).

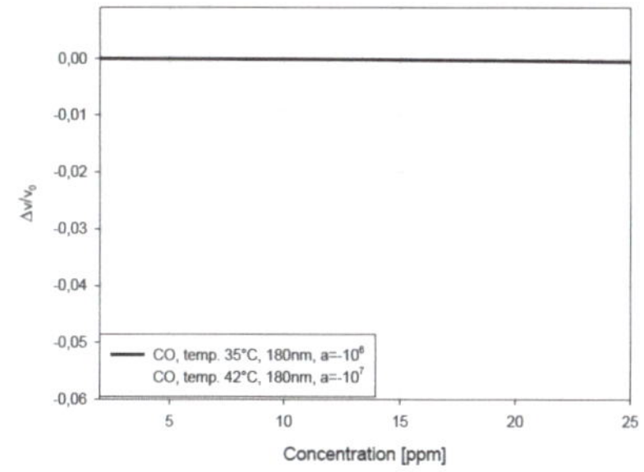

Figure 17.
Changes of velocity (relative) vs. concentration from 2 to 25 ppm in synthetic air, Nafion+polyaniline at temperatures 35°C and 42°C. assumed in the analysis: H = 180 nm (thickness), sensitivity coefficient $a = -10^6$ *ppm*$^{-1}$ *(T = 35°C),* $a = -10^7$ *ppm*$^{-1}$ *(T = 42°C),* $\sigma_s = 2 \times 10^{-12}$ *S, M = 28 g/Mol (CO). Numerical results.*

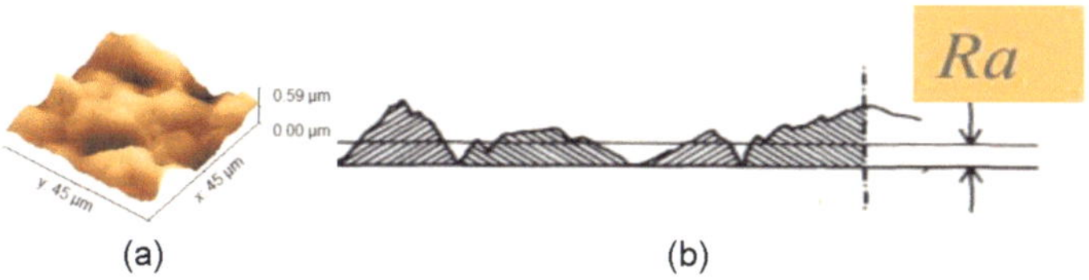

Figure 18.
The polymer layer (RR)-P3HT surface 3D-AFM view (a) graphic illustration of the average profile radius (b).

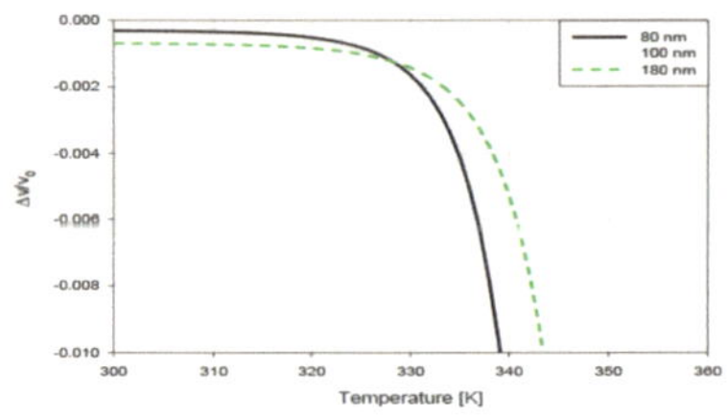

Figure 19.
Changes of velocity (relative) vs. temperature from 300 K to 344 K, Nafion+polyaniline, assumed in the analysis: $A = -1\ ppm^{-1}$, $\sigma_s = 2 \times 10^{-12}$ *S, M = 28 g/Mol (CO), concentration 5 ppm, thickness 80, 100, 180 nm. Numerical results.*

The sensitivity of sensing structures to external gas atmospheres depends on the temperature of the structure. The problem of the structure temperature is also important in the aspect of "detoxifying" its electrical properties after measurements. Our previous research [2] shows that detoxification is much more effective at elevated temperatures. On the other hand, however, the elevated temperature affects the mechanical degradation of the structure and causes irreversible changes in its physicochemical properties. This problem is particularly important when the sensing layers are made of organic semiconductors. Polyaniline and Nafion are organic semiconductors. In order not to destroy the examined structures, the measurements were made at relatively low temperatures: 35°C (308 K) and 42°C (315 K).

Analyzes for the sarin simulator—DMMP gas was also performed. The roughness of the polymer [25] detector layer for the average height of the layer profile from 60 nm to 960 nm (max. average height of the layer profile) theoretically and numerically was performed. The thickness using the atomic force microscope (AFM) profile analysis was estimated too. The acoustoelectric interaction of the surface wave with charge carriers distributed in the detector layer according to the profile resulting from the diffusion of gas molecules from the surrounding atmosphere was analyzed numerically in the Python programming environment, using the expression (1)–(5) [10]. The results of the numerical analysis can be the basis for the optimization of the layer parameters in terms of the maximum sensitivity of the detector. **Figures 20**, **21**, and **22–24** show exemplary results of the numerical analysis.

The best results in the range of 60–960 nm thicknesses have been achieved. In **Figure 20** analysis for a polymer (RR)-P3HT was performed. For the assumed DMMP gas ambient parameters (temperature, concentration 3 ppm) and layer parameters

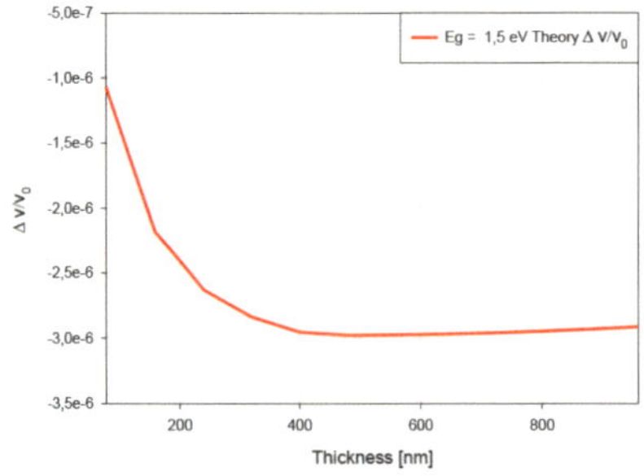

Figure 20.
Changes of velocity (relative) vs. thickness, from 80 nm to 960 nm, layer (RR)-P3HT, gas DMMP: $\sigma_s = 5 \times 10^{-4}$ *S, M = 124.08 g/Mol,* $K_2/2$ *(quartz) =0.09%, C = 3 ppm.*

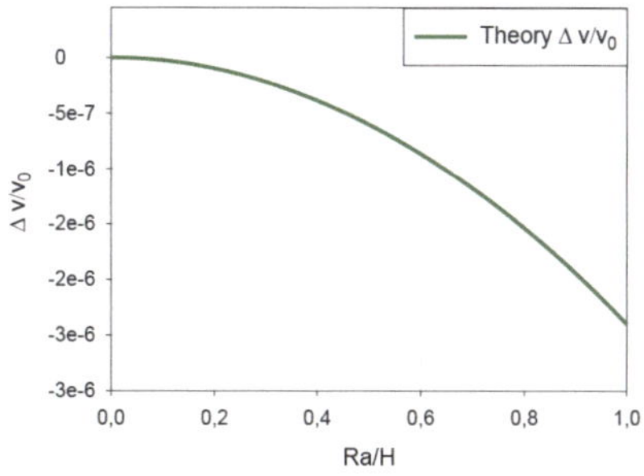

Figure 21.
Changes of velocity (relative) vs. roughness, temp. 305 K (32°C), (RR)-P3HT, a = 1 ppm^{-1}, $\sigma_s = 5 \times 10^{-4}$ S, $D_K = 10^6 nm^2 s^{-1}$, $E_g = 2.7$ eV, M = 124.08 g/Mol, DMMP concentration: 2 ppm, thickness 500 nm (H), $K_2/2$ (quartz) =0.09%, measurement AFM $R_a = 76$ nm.

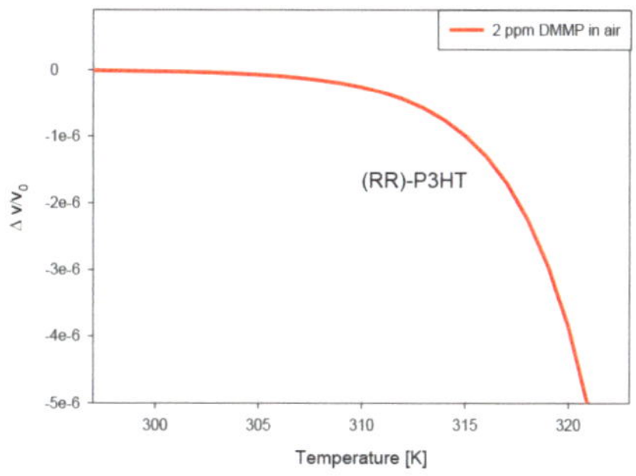

Figure 22.
Temperature range: 293–322 K (20–49°C)—Curve (theoretical numerical calculations): $\sigma_s = 5 \times 10^{-4}$ S, M = 124.08 g/Mol, $K_2/2$ (quartz) =0.09%, C = 2 ppm.

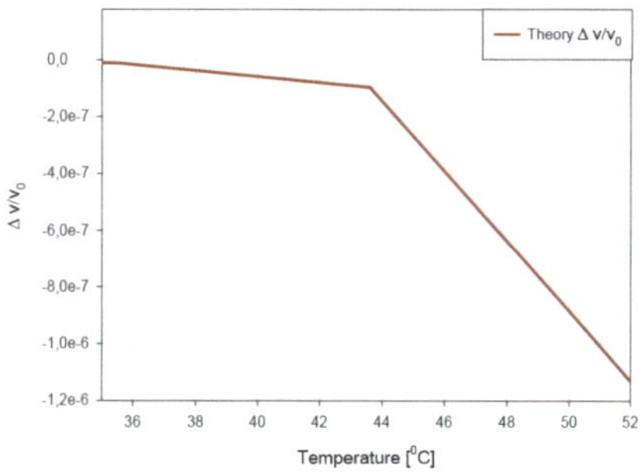

Figure 23.
Curve (theoretical numerical calculations) for parameters: $\sigma_s = 5 \times 10^{-4}$ S, M = 124.08 g/Mol, $K^2/2$ (quartz) =0.09%, C = 2 ppm.

(sensitivity, conductivity, diffusion parameters, substrate—quartz), the graph shows the value of about 3×10^{-6} for a layer with a thickness of 500 nm. The changes resulting from the numerical model are at a relative level of changes of the order of 10^{-6}. The results shown in **Figure 20** show that there is an optimal thickness of the detector layer for which the acoustoelectric impact (change in the velocity of the acoustic wave) is the greatest. Qualitatively, the existence of an optimal layer thickness was confirmed empirically in Ref. [7, 8].

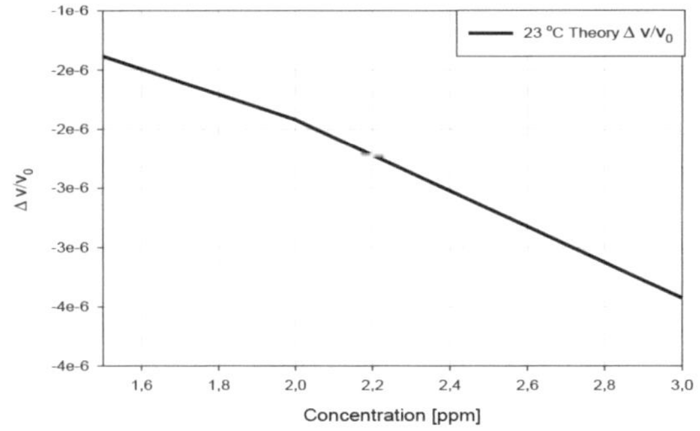

Figure 24.
Changes (relative) of velocity vs. concentration, temp. 305 K (32°C), (RR)-P3HT, $a = -1.75\ ppm^{-1}$, $\sigma_s = 5 \times 10^{-4}$ *S,* $DK = 10^6 nm^2 s^{-1}$, $E_g = 2$ *eV,* $M = 124.08$ *g/Mol, gas DMMP, thickness 500 nm,* $K_2/2$ *(quartz) = 0.09%.*

The thickness of the layer affects the response of the detector layer, also in this case. In our case, we identified the height of the layer with the average radius of the detector layer profile. This is a particularly favorable circumstance because in the perturbation theory, the concept of surface conductivity is used. The following graph was obtained by equating the average radius of the R_a layer with the thickness of the layer $\sigma_s = \sigma R_a$. This fact gave our considerations a further direction enabling the development of a mathematical model to take into consideration roughness. In this case, most of the volume of the layer is close to the surface of the piezoelectric layer of the quartz base. The diffusion of larger molecules is not difficult, however, in layers of greater thicknesses above about 650 nm, the interaction is not as strong as in smaller ones, eg. 100 nm. The electrical conductivity of the deeper parts of the layer changes, and their contribution to the detector response is significant. In the case of detectors based on polymer layers, the shape of the surface and its roughness are significant. The above characteristics of the detector response depending on the average radius of the layer's profile were shown (**Figure 18**). The average radius of the R_a layer profile by the height of the H-layer for the better presentation was normed (R_a/H—**Figure 21**). In the case of the average profile radius R_a of 0, the layer "disappears", boundary conditions correspond to the lack of interaction of the analyte with the detector layer. In the case when the average radius of the R_a—layer profile reaches the height of the maximum H. This is a theoretical case when we have an ideal detector layer in the image of a "rectangular"—and the surface that is not rough—a theoretical case. Therefore, the roughness has decisive for the response of the DMMP gas detector based on the optic layer using (RR)-P3HT and this aspect will be developed in further studies. In the case of DMMP, the optimal layer thickness is approx. 500 nm. These analyzes were carried out for other analytes, such as: H_2, CO_2, NO_2, NH_3. Numerical NNA analysis also showed the existence of an optimal thickness at which the interaction is greatest. The optimal thickness for $LiNbO_3$ (piezoelectric) layer and interaction with gases H_2, CO_2, NO_2, NH_3 was approx. 100 nm [9]. As a rule, the diffusion of larger molecules is difficult, additional in layers of higher thicknesses, the entire volume of the layer is not saturated. However, in the case of polymers, it is the other way round. The electrical conductivity of the deeper parts of the layer also changes, so their contribution to the detector response is significant. This state describes the solution of the general diffusion equation for polymers [26].

The results of the experiment to the theory were compared. **Figure 22** shows theoretical numerical analysis in the temperature range of about 293–322 K. Below a comparison of numerical calculations based on the assumed layer model with empirical measurements was made. Good consistency of row magnitude was achieved. The results of the experiment coincide with the theory and are convergent. The calculations for a DMMP concentration of 2 ppm were made.

In **Figure 24** the theoretical dependence of the detector response depending on the concentration in the range from 1 to 3 ppm was shown. The results of empirical research concern the concentration of 1.5–3 ppm and at 32°C were made. All tests on a piezoelectric substrate made of quartz were done.

5. Summary

In research layers (PANI) for different thicknesses: 100, 180 nm, Nafion approx. 300 nm were prepared. The main target was to research the interaction of carbon monoxide and a layer of PANI change were measured. Influence the temperature on the bilayer structure of Nafion+Polyaniline at T = 35° C and T = 42° C were examined. Visible effects were observed at low CO concentrations—5 ppm. During the experiment, representative samples of the PANI with a Nafion layer thicknesses of 100 and 180 nm were investigated. Very interesting properties were examined at layer 180 nm depending on the interaction temperature. Very clearly and a sharp rise detection of carbon monoxide properties as a result of the growth of the temperature of about 7°C (temperature change from 35°C to 42°C) was observed. Exactly ambient temperatures were measured. Empirical results with layer basis on PANI + Nafion (100, 180 nm) compared with Polypyrrole detector layer (thickness $\sim$ 80 nm). It is useful to measure a higher level of the concentration of CO above 25 ppm. Extensive numerical analyses on CO of the SAW response parameter like: thickness, concentration, and temperature at low concentrations (5 ppm) were conducted. Numerical researches were performed in a steady state. As shown above, PANI and PPy nano-layers have a lot of valuable physical properties useful for applications in various fields of science. These materials are used in medical and biological applications, eg. for the detection of both chemicals, e.g. the selected metal cations, lactic acid as well as biological samples—tumor cells. Polyaniline nano-layers can develop a detector that can detect lactic acid in the range of physiological concentrations informing about the different types of diseases. Empirical studies can observe specific changes in the electrical conductivity of micro and PANI nanolayers in contact with tumor cells and healthy [13].

Detector studies of photoconductive (RR)-P3HT were performed as a potential material for detecting trace amounts of DMMP compound vapors (2 ppm) in air using the SAW method. Polymer (RR)-P3HT possesses significant sensitivity to trace amounts of DMMP, only when additional white light was used. Increasing the light flux also causes the sample temperature to rise and to obtain larger frequency variations (130–300 Hz) for the same concentration of 2 ppm DMMP in the air. Estimated response and regeneration times for this DMMP concentration are respectively 10–20 sec. and 7 min [1]. Extensive NNA of the SAW detector response depending on parameters: concentration, roughness, temperature, and thickness were conducted. Numerical research was performed in a steady state. NNA was performed using proprietary software written in Python. The layer thickness decides the maximum range of the detected gas concentration [10].

Also, the choice of the temperature of the polymer detector is important—the optimal work temperature depends on the type of the detector layer and its roughness. The change in temperature allows to determine the point of detector work, and has also an impact on the speed of its response and the recovery time of the sensing properties of the detector layer [9]. Empirical results of a SAW detector with (RR)-P3HT were achieved for three thicknesses (~100,~350,~550 nm) of the layers. The main target was to verify empirically the analytical model of the SAW gas detector affected by: 2, 3 ppm DMMP in the air. Empirical results confirmed the usefulness of the elaborated analytical model for the investigation of the SAW detector in the design stage. In particular, the influence of the concentration and thickness (existing optimal thickness) was confirmed. The essential parameter of the polymer detector is the roughness of the sensing layer—under investigation.

The results of numerical acoustoelectric analysis of the SAW detector investigations confirmed theoretical were showed. Different sources of light (different of wavelengths light LED, see **Figure 13**) were selected and different wavelengths (histogram) were checked.

The study has shown that the parameters of the detector layer for the SAW detector should be individually adjusted, according to the type of the detected gas and the applied sensing layer. The exposure time of polymer layers was also empirically selected for about 10 min for each current value of 100, 200, 300 mA and the detector layer was illuminated with LED light of different wavelengths and different light energy. From the tests, the best results for yellow light: 76.6–87.4 [lm] and wavelength = 560 nm were obtained. The author's main goal was to numerically analyze phenomena based on empirical investigation and to create a predictive model that would show the tendency of changes before they appear. Extensive numerical analyses of the SAW detector response depending on parameters like: concentration, roughness, temperature, and thickness were conducted. Numerical investigation were performed in the steady state conditions in Python.

Acknowledgements

Authors would like to thank the Presidents: Wojciech Wajda WASKO SA (innovation technology and research Enterprise) and Michał Mental ENTE Ltd. (the newest electronic factory) for the possibility of work: theoretical, analytical, numerical, and investigation works as well as participation in R & D works financed from the NCBiR grants. The authors of the paper thank Paulina Powroźnik (intensive research), Anna Kaźmierczak-Bałata (morphology), Maciej Setkiewicz (performance sensing layer and technology), Mirosław Magnuski (electronic design of appliance high frequency and patent) Monika Hejczyk—President Foundation—The Academy of Creative Development (statut support for science) for his help in the experimental investigations.

Author details

Tomasz Hejczyk[1], Jarosław Wrotniak[2] and Wiesław Jakubik[3]*

1 The Academy of Creative Development—The Foundation, Marklowice, Poland

2 Institute of Electronics, Silesian University of Technology, Gliwice, Poland

3 Institute of Physics CSE, Silesian University of Technology, Gliwice, Poland

*Address all correspondence to: wieslaw.jakubik@polsl.pl

References

[1] Wrotniak J, Jakubik W, Powroźnik P, Stolarczyk A, Magnuski M. Akustyczne badania polimeru typu RR-P3HT do wykrywania DMMP w powietrzu (in polish). Przegląd Elektrotechniczny. 2018;**94**(6):70-73. DOI: 10.15199/48.2018.06.12

[2] Hejczyk T, Pustelny T. Analysis of the saw system with the PANI + Nafion sensing structure for detection of low concentration carbon monoxide. Archives of Acoustics. 2020;**45**(4):681-686. DOI: 10.24425/aoa.2020.135274

[3] Matsumaga N, Sakai G, Shimanoe K, Yamazoe N. Diffusion equation-based study of thin film semiconductor gas sensor-response transient. Sensors and Actuators B. 2001;**83**:216-221

[4] Matsumaga N, Sakai G, Shimanoe K, Yamazoe N. Formulation of gas diffusion dynamics for thin film semiconductor gas sensor based on simple reaction-diffusion equation. Sensors and Actuators B. 2003;**96**:226-233

[5] Hejczyk T, Pustelny T, Wszołek B, Jakubik W. Numerical analysis of sensitivity of SAW structure to the effect of toxic gases. Archives of Acoustics. 2016;**41**(4):747-755. DOI: 10.1515/aoa-2016-0072

[6] Pustelny B, Pustelny T. Transverse acoustoelectric effect applying in surface study of GaP: Te(111). Acta Physica Polonica A. 2009;**116**(3):383-384. DOI: 10.12693/APhysPolA.116.383

[7] Hejczyk T, Urbanczyk M, Pustelny T, Jakubik W. Numerical and experimental analysis of the response of a SAW structure with WO_3 layers an action of carbon monoxide. Archives of Acoustics. 2015;**40**(1):19-24. DOI: 10.1515/aoa-2015-0003

[8] Hejczyk T, Wrotniak J, Magnuski M, Jakubik W. Experimental and numerical acoustoelectric investigation of the new SAW structure with (RR)-P3HT polymer in DMMP detection. Archives of Acoustics. 2021;**46**(2):313-322. DOI: 10.24425/aoa.2021.136585

[9] Hejczyk T, Urbańczyk M. Numerical optimization of structures SAW gas sensors. Acta Physica Polonica A. 2013;**124**(3):432-435. DOI: 10.12693/APhysPolA.124.432

[10] Urbańczyk M. Sensors with Surface Acoustic Wave—Monograph. Vol. 213. Gliwice (in Polish): SUT; 2011

[11] Pustelny T, Opilski A, Pusteny B. Determination of some kinetic parameters of fast surface states in silicon single crystals by means of surface acoustic wave method. Acta Ph ysica Polonica A. 2008;**114**(6A):A183-A190

[12] Sakiewicz P, Nowosielski R, Babilas R, Gąsiorek D, Pawlak M. FEM simulation of ductility minimum temperature (DMT) phenomenon in CuNi25 alloy. Journal of Achievements in Materials and Manufacturing Engineering. 2013;**61**(2):274-280

[13] Michalska A. Polyaniline Micro- and Nanostructures for Biomedical Applications. PhD Thesis (in Polish); 2011

[14] Hejczyk T, Kamiński G, Ogaza R. Patent no PL 229696 B1, Application of the Hybrid Sensor with Acoustic Surface Wave [in Polish]. 2018. pp. 2-5

[15] Powroźnik P, Jakubik W, Kaźmierczak-Bałata A. Detection of organophosphorus (DMMP) vapour using phthalocyanine-palladium bilayer

structures, EUROSENSORS. Procedia Engineering. 2015;**120**:368-371. DOI: 10.1016/j.proeng.2015.08.640

[16] Magnuski M, Wrotniak J. Patent no PL 230526 B1, System for detecting chemical compounds in gaseous atmospheres, with a sensor using surface acoustic waves (SAW). Polis. 2018:2-4. P.419823

[17] Jasek K, Miluski W, Pasternak M. New approach to saw gas sensors array response measurement. Acta Physica Polonica A. 2011;**120**(4):639-641. DOI: 10.12693/APhysPolA.120.639

[18] Pustelny T, Procek M, Maciak E, Stolarczyk A, Drewniak S, Urbanczyk M, et al. Gas sensors based on nanostructuures of semiconductor ZnO and TiO2. Bulletin of the Polish Academy of Sciences: Technical Sciences. 2012;**60**(4):853-859

[19] Kawalec A, Pasternak M. Microwave humidity sensor based on surface acoustic wave resonator with nafion layer. IEEE Transactions on Instrumentation and Measurement. 2008;**9**(57):2019-2023

[20] Kawalec A, Pasternak M, Jasek K. Measurements results of SAW humidity sensor with nafion layer. European Physical Journal: Special Topics. 2008; **154**:121-126

[21] National Research Council. Acute Exposure Guideline Levels for Selected Airbone Chemicals. Washington, DC: The National Academies Press; 2003

[22] Yoo R, Kim J, Song MJ, Lee W, Noh JS. Nano-composite sensors composed of single-walled carbon nanotubes and polyaniline for the detection of a nerve agent simulant gas. Sensors & Actuators, B: Chemical. 2015; **209**:444-448. DOI: 10.1016/j.snb.2014. 11.137

[23] Auld BA. Acoustic Fields and Waves. Vol. 2. New York: Wiley; 1973. pp. 271-290

[24] Long Y, Wang Y, Du X, Cheng L, Wu P, Jiang Y. The different sensitive behaviors of a hydrogen-bond acidic polymer-coated SAW sensor for chemical warfare agents and their simulants. Sensors. 2015;**15**:18302-18314. DOI: 10.3390/s150818302

[25] Lee HJ, Park KK, Kupnik M, Oralkan Ö, Khuri-Yakub BT. Chemical vapor detection using a capacitive micromachined ultrasonic transducer. Analytical Chemistry. 2011;**83**(24): 934-932. DOI: 10.1021/ac201626b

[26] Crank J. The Mathematics of Diffusion. Oxford: Clarendon Press; 1975. pp. 259-262